DISCOVERIES OF THE CENSUS OF MARINE LIFE
Making Ocean Life Count

Over the 10-year course of the recently completed Census of Marine Life, a global network of researchers in more than 80 nations has collaborated to improve our understanding of marine biodiversity – past, present, and future.

Providing insight into this remarkable project, this book explains the rationale behind the Census and highlights some of its most important and dramatic findings, illustrated with full-color photographs throughout. It explores how new technologies and partnerships have contributed to greater knowledge of marine life, from unknown species and habitats, to migration routes and distribution patterns, and to a better appreciation of how the oceans are changing. Looking to the future, it identifies what needs to be done to close the remaining gaps in our knowledge, and provides information that will enable us to manage resources more effectively, conserve diversity, reverse habitat losses, and respond to global climate change.

PAUL SNELGROVE is a Professor in Memorial University of Newfoundland's Ocean Sciences Centre and Biology Department. He chaired the Synthesis Group of the Census of Marine Life that has overseen the final phase of the program. He is now Director of the NSERC Canadian Healthy Oceans Network, a research collaboration of 65 marine scientists from coast to coast in Canada that continues to census ocean life.

DISCOVERIES OF THE
CENSUS OF MARINE LIFE

Making Ocean Life Count

PAUL V.R. SNELGROVE

Memorial University of Newfoundland, St. John's, Canada

CAMBRIDGE
UNIVERSITY PRESS

CAMBRIDGE UNIVERSITY PRESS
Cambridge, New York, Melbourne, Madrid, Cape Town, Singapore,
São Paulo, Delhi, Dubai, Tokyo, Mexico City

Cambridge University Press
The Edinburgh Building, Cambridge CB2 8RU, UK

Published in the United States of America by Cambridge University Press, New York

www.cambridge.org
Information on this title: www.cambridge.org/9781107000131

First published 2010

Printed in the United Kingdom at the University Press, Cambridge

A catalogue record for this publication is available from the British Library

Library of Congress Cataloging-in-Publication Data

Snelgrove, Paul V. R., 1962–
Discoveries of the Census of Marine Life : Making Ocean Life Count / Paul V. R. Snelgrove.
    p.   cm.
  ISBN 978-1-107-00013-1 (Hardback) – ISBN 978-0-521-16512-9 (Pbk.)
  1. Marine biodiversity.   2. Census of Marine Life (Program)   3. Marine ecology.
4. Marine biology–Research–Methodology.   5. Ocean.   I. Title.
  QH91.8.B6S64 2010
  578.77–dc22

                                                            2010028669

ISBN 978-1-107-00013-1 Hardback
ISBN 978-0-521-16512-9 Paperback

Additional resources for this publication at www.cambridge.org/9781107000131

For Fred – Scholar, mentor, friend

CONTENTS

FOREWORD

We went first to the Moon

Life on Earth originated in the margins of the primordial ocean and for billions of years evolved in this aquatic milieu. Life originated on Earth because it has the ocean: water in liquid state on its surface. Although curiosity has pushed humanity to search for life in outer space, we still know no other planet with life in the universe. One Planet, one Ocean.

Unquestionably, from a planetary perspective, the ocean is a thin layer of fluid that plays an essential role in making the planet livable. The ocean is to Earth thinner than the skin to an apple.[1] However, to our human scale and resources, the oceans represent a vast tri-dimensional space, opaque to our vision and full of unknowns. In the 1960s, in the middle of the Cold War and the race for space exploration, a campaign called for conquest of "the inner frontier," to mount a major research and technology effort to explore the ocean interior. As we all know now, we went first to the Moon. We haven't extracted a single gram of food from the Moon, but in the last 60 years we have extracted over 3,500 million tons of fish from the ocean and roughly half of our natural gas and oil come today from the ocean floor.

But what we always fail to understand, beyond these immediate and selfish uses, is that the ocean provides essential ecological services to all humanity, making life possible on our planet: *the ocean is the ultimate global commons,*

---

[1] On average the radius of the planet is 6,371 kilometers, the ocean on average is 3,733 meters deep, i.e. a thickness of 0.058% or 6 ten-thousandths of the radius.

something that belongs to all of us. One example: marine photosynthesis produces annually 36 billion tons of oxygen, estimated to equal 70% of the oxygen in the atmosphere. I cannot think of a more fundamental reason to assert that every form of life on Earth has a stake in the health of the ocean. Humanity, mastering the technological power to disrupt its equilibrium, is especially responsible for its health.

The ocean under the pressure of numbers ... numbers of humans, that is

Many human uses of the ocean have secondary effects that adversely impact the stability of natural processes in the ocean. Alarming signs alert us that the integrity of several natural systems that provide basic ecological services to humanity is threatened. Evidence accumulates to demonstrate that the management systems that we use do not suffice to guarantee the sustainability of living marine resources.

Destruction of critical coastal habitats is alarming, as human populations fill the shores. Destruction of deep-ocean habitats is significant due to the secondary effect of fish trawling. The frequency and size of dead zones increase due to the exhaustion of oxygen by the arrival of vast quantities of chemicals used by, or originating in, industry, agriculture, and animal husbandry, and transported by rivers into the ocean. Satellites and sailors detect massive accumulation of plastic in the central gyres of the Pacific Ocean.

Absorbing millions of tons of CO_2 every year – roughly one-third of total annual emissions – the ocean has already spared us from catastrophic climate change. But in doing so, its own intrinsic balances are altered: the ocean is becoming more acidic and has taken the largest fraction of the additional heat generated by anthropogenic greenhouse gases, something that might eventually alter the normal patterns of ocean circulation essential for keeping the absorbed CO_2 from reuniting with the atmosphere for long periods, buying us time for finding solutions to climate change.

We have an incomplete and piecemeal picture of what is happening to the ocean and an urgent need to generate the compelling evidence that should force us to adopt corrective policies at the highest level possible. Too much is at stake to follow the path of least resistance. The book that you have in your hands is the first digest synthesizing a unique scientific program designed to move the boundary of the unknown in the ocean: The Census of Marine Life, catalyzed by support from the Alfred P. Sloan Foundation and composed of 17 different international projects that in 10 years mobilized more than 2,700 researchers in 500$^+$ research expeditions, added over a thousand new species to science (and counting), put together nearly 28 million records of individual ocean specimens, and produced thousands of contributions to the scientific literature.

Abnormal science

Normal science builds upon the many contributions that researchers make when setting for themselves a research goal that eventually results in a published paper. Each step tries to answer the immediate knowledge gap in a logical sequence of analysis of a single phenomenon. Apparently there is no *a priori* plan, but collectively these individual contributions build the edifice of science, and push back the boundary of the unknown.

The Census of Marine Life is a different intellectual enterprise. Disregarding many objections from *Mainstream Road*, the leaders of the initiative used a metaphor to rally the interest of the relevant scientific community: to conduct a *Census* of marine life, an impossible task *sensu strictu*. By choosing an extremely broad subject, the living ocean, and setting a research vector, or direction, to count and account for the living in the ocean, the founders were able to form a community of researchers with quite disparate research interests and objectives, to weave a delicate fabric of research topics that brought together the main ingredients of scientific discovery: deploying new technologies, poking through disciplinary boundaries, transporting knowledge produced in one field to another, attacking simultaneously the small and the large and the extremely large scales usually unavailable to single teams of scientists. Using as an epistemic Occam's razor the distinction between the known, the unknown, and the unknowable, they collectively and systematically selected a limited number of bets to maximize results. This book demonstrates unreservedly their success.

Around the living ocean in 10 chapters

Modern scholastics divide the study of the ocean in the physical, chemical, and biological oceanographies; marine biology, concentrating more on the organisms themselves and currently flourishing thanks to genomic techniques;

and marine geology and geophysics, plus all the engineering subsumed in the applied ocean sciences. What this scholastic division misses is that the ocean itself is alive. I am not falling into a mystic lapse suggesting a Spencerian superorganism. The discovery both of diverse chemosynthetic biological communities of hydrothermal vents and deep-ocean seeps and of the rich and abundant microbial life in the upper 100 meters of sedimentary ocean bottom are new facts changing our collective perception of the ocean. Without understanding life in the ocean we will never understand the complex system that the ocean is. Life is immediate to chemical and biological oceanography and geology. The ultimate equilibrium of climate on the planet most likely will be biologically set. Following a long and venerable tradition, the Census went out with new tools *to study patterns within this living ocean.* This book reports back that this choice was extremely fortunate and successful, turning up everywhere discoveries, as these fascinating pages reveal.

I will not summarize here the content of each chapter of this digest. I will idiosyncratically choose certain highlights.

Animals that are strong swimmers may move distances of hundreds or even thousands of kilometers in pursuit of mates, food, suitable temperatures, and oxygen that enhance their survival, growth, and reproductive success. On the technology side, the Census developed new electronic tags for organisms that were massively used. Within the Tagging of Pacific Predators (TOPP) project some tags recorded physiological functions, most were capable of determining geographical position through GPS, and many transmitted data through satellite, allowing the tracking of animal movement across the ocean. This enabled the identification of "hot spots," "cold spots," "highways," and "truck stops" of many different types of animals. These are ocean regions where they feed, reproduce, or correspond to preferred migration routes. This information allows us fascinating glimpses into how animals, other than humans, use the marine environment. It provides unique and highly applicable information for the protection and management of the ocean.

Still on the technology side, within the Pacific Ocean Shelf Tracking Project (POST) the development of large "curtains" of sensors enabled the precise counting of individual organisms while they massively migrate in the ocean. This is being used and applied to monitor salmon populations in the North Pacific and is already informing management decisions.

The Census benefited from the fast development of molecular biology and genomics in recent years, the precise reading and identification of genetic material in organisms. The "barcode of life" methodology broadly applied as part of the Census of Marine Zooplankton (CMarZ) project can rapidly reveal whether two organisms belong to the same species. A barcode of life is a short DNA sequence from a uniform locality on the genome that can provide a true molecular ID card for each marine species. This allows major steps forward in elucidating the presence of many cryptic species in the ocean, by distinguishing species that superficially resemble one another, and joining specimens that vary in appearance from one region to the next. CMarZ has targeted potential biodiversity hot spots throughout the world, including poorly known regions such as Southeast Asia, the polar oceans, and the water column below 5,000 meters. Another related molecular technique, pyrosequencing, was widely used by the International Census of Marine Microbes (ICoMM) project, as they built a completely new picture for marine microbial diversity and abundance, and the role microbes play in the global ocean.

A single liter of seawater can contain more than one billion microbes.[2] Marine microbes account for perhaps half of the primary production that fuels all life on Earth and they control the global cycling of nitrogen, sulfur, iron, and manganese. Without microbes, life on Earth could not exist.

The Arctic Ocean Diversity (ArcOD) project and the Census of Antarctic Marine Life (CAML) both participated in barcoding. Using similar techniques to those championed by CMarZ and ICoMM, polar microbiologists discovered 1,500 kinds of Arctic *bacteria* and 700 kinds of *archaea*. CAML has added over 11,000 barcode sequences for Antarctic species from their collections. The Census is working with the Marine Barcode of Life (MarBOL) to compile a marine library and expects to have accumulated reference codes for 50,000 species by the end of 2010. Researchers active in the Census of Coral Reef Ecosystems (CReefs) are working to apply variants of environmental genomics involving mass sequencing developed for microbes to assist in coral reef taxonomy.

I cannot finalize this review without mentioning the History of Marine Animal Populations (HMAP) and the Future of Marine Animal Populations (FMAP) projects, which both studied changing oceans. Employing a wide variety of historic, anthropological, and natural-science methods and techniques, HMAP made the most serious effort to date to reconstruct a vision of life in the ocean before massive human interference. HMAP has produced sobering baselines that should add depth and effectiveness to management

---

[2] Approximately 100,000,000,000,000,000,000,000,000,000 (or 10^{29}) total bacteria in the global ocean and about 20,000 operational taxonomic units (a proxy for species) in a typical liter of seawater.

decisions. FMAP has continued that timeline forward to produce new global views of diversity, distribution, and abundance that illustrate the current scope of human impact. These and all the other Census projects contribute data to the Ocean Biogeographic Information System (OBIS), the Census biodiversity data legacy.

To conclude I want to highlight the use that the Census made of innovative sampling techniques to access remote and inaccessible areas of Planet Ocean. The five Census projects looking specifically into the deep ocean, down the continental slopes, through vents and seeps, across the abyssal plains, and up over seamounts and the Mid-Atlantic Ridge, used new submersible tools, towed cameras, remotely operated vehicles (ROVs), autonomous underwater vehicles (AUVs), and manned submersibles.

Da capo a fine

In a fortunate coincidence, Chapter 10: "Planet Ocean beyond 2010," brings us full circle to the question of why humanity went first to the Moon: "Already satellites continuously orbit the Earth, collecting imagery for myriad applications. Soon autonomous underwater vehicles (AUVs) may move across the seafloor like underwater satellites. AUVs today run well-defined, pre-programmed missions (…) The next generation AUVs may visit underwater docking stations to recharge batteries and download data." This extremely powerful vision is also a declared need of the international community. The World Summit on Sustainable Development in 2002 decided to keep the oceans under permanent review via global and integrated assessments of the state of ocean processes. This initiative for a world ocean assessment is the most comprehensive yet undertaken by the United Nations system to improve ocean governance. Implementing it will make full use of the baseline 2000–2010 established by the Census. Maintaining it through time will require a permanent program of observations of the living oceans that complements what we now have for the physics of the ocean and climate. As this digest abundantly demonstrates, the Census of Marine Life did its part.

PATRICIO A. BERNAL
Executive Secretary (1998–2009)
Intergovernmental Oceanographic Commission
Paris, March 2010

PREFACE

The Census of Marine Life has been the opportunity of a lifetime to travel
the world and meet wonderful scientists who each bring their own passion,
toolbox, and diverse view of the ocean. The Census is about the thousands of
scientists who have bounced around on ships, slogged through samples, and
spent countless hours hunched over their computers trying to bring clarity to
an opaque ocean. The project leaders have kindly shared stories, manuscripts,
imagery, and ideas that are the core of this book. Their work builds on
that of many other talented scientists around the world.

The goal of this book is to bring the excitement of the Census and its
findings to as broad an audience as possible. This book encompasses many
hundreds of science papers, but to improve readability, I simply include a
list of those by chapter rather than within the chapter text itself. An online
"educators" version of the book, available through Cambridge University
Press (www.cambridge.org/9781107000131), includes reference citations
within the text so anyone interested can link specific information to its source.

The task of trying to corral many different Census activities into
coherent outputs has been a collective adventure with the Census Synthesis
Group that I chair, which includes Jesse Ausubel, Darlene Trew Crist, Michele
DuRand, Fred Grassle, Pat Halpin, Sara Hickox, Patricia Miloslavich, Ron
O'Dor, Myriam Sibuet, Edward Vanden Berghe, Boris Worm, and Kristen
Yarincik. Generous funding from the Alfred P. Sloan Foundation allowed
us to focus on ideas and outputs that span this book and beyond to capture
different aspects of the Census. We have benefited enormously from working
with the Census Scientific Steering Committee led first by Fred Grassle

and then by Ian Poiner. The Secretariat managed by Kristen Yarincik has done a spectacular job orchestrating everything Census.

The hospitality of Heidi Sosik, Judy McDowell, and the Biology Department at Woods Hole Oceanographic Institution created a great opportunity to research much of this book. That time was made possible by the support and flexibility of my colleagues at Memorial University, Ian Fleming, Paul Marino, Garth Fletcher and Mark Abrahams. The Canadian Healthy Oceans Network team most especially Joan Atkinson, made a confluence of opportunities work, and covered for me when needed.

My graduate students in Newfoundland were wonderfully patient and helpful in their spare time, particularly Krista Baker who did an outstanding job formatting the many references that Ashlee Lillis helped hunt down. Ryan Stanley kept the lab afloat in my absence.

Feedback and ideas on early chapters from Ron O'Dor, Patricia Miloslavich, and Darlene Trew Crist helped to set the tone for the book. Reviews by Michael Sinclair, Serge Garcia, Vera Alexander, and an editorial "haircut" by Jesse Ausubel and Paul Waggoner greatly improved the content. Excellent support from the Mapping and Visualization Team led by Pat Halpin and the Education and Outreach Team led by Sara Hickox were critical in producing the figures and imagery in this book, often at short notice. Frank Baker's efforts to secure photo credits were also a tremendous help.

Michele DuRand was invaluable, spending many hours proofing, editing, and advising. Darlene Trew Crist provided tremendous counsel on many important details. The patience and flexibility of Martin Griffiths and Lynette Talbot at Cambridge University Press helped greatly.

Family members lent much support and humor, most importantly the companionship and encouragement of my wife along this long voyage of discovery.

PART ONE

The unknown: why a Census?

1

Planet Ocean

We are a species that breathes air rather than water, which biases our view of the watery Planet Earth. We might better call a globe with 71% water cover and more than 99% of its living biosphere in marine waters Planet Ocean, as noted years ago by the writer Arthur C. Clarke (Figure 1.1). We are far less familiar with the ocean than the land. Even for scientists working in the ocean for all of their lives there are surprises. Shocked to see a recent photograph of a clam named *Pholadomya candida* that was thought to have been extinct since the late 1800s, a Colombian scientist spent many days diving and searching where the photograph had been taken. Just before his scuba tank ran dry one day, he found a living specimen a little deeper and in colder water than expected, rescuing it, at least temporarily, from the list of extinct species. Deep-water samples my research team recently

Figure 1.1 The predominance of water on Planet Ocean.
Census scientists explored marine life in all realms of the world's oceans, visualized here on a Blue Marble image.

collected from Eastern Canada were examined by a world authority on marine worms, who was surprised that half of the species had never been named. This novelty appears to be the norm for many deep-water environments, where as much as 85% of the species in a sample can be new to science.

These anecdotes illustrate how little we know about Planet Ocean and what lives in it. The ocean is changing, and it is changing before we have really appreciated all that it now contains. What lives in the ocean? Where do they live? How many are there? In three words, what is their diversity, distribution, and abundance? We are further from answering these basic questions than even most scientists realize.

To address these three major questions, marine experts from around the world banded together in the year 2000 in an international initiative called the Census of Marine Life. In the decade of discovery that followed, as this book summarizes, the Census and others focused new "binoculars" of technology to look into the ocean and propelled forward the understanding of diversity, distribution, and abundance of marine life. Through the chapters that follow, we will visit the different areas of the ocean and use these new "binoculars" to see and understand single-celled microbes and whales, and everything in between, in ways not possible only a decade ago. In short, Planet Ocean is coming into focus, and is becoming more transparent.

The array of marine life and how they live varies immensely. The size of marine organisms ranges a hundred million million million fold, from drifting bacteria through blue whales. From the smallest to the largest organisms, from the shortest to the longest lived, from the slowest to the fastest, and from the drab to the flamboyant, living organisms have developed an amazing range of strategies to survive. Some species, such as bacteria, may only live for hours, but they can also replicate in that time and produce multiple generations in the time it takes us to get a good night's sleep. At the other extreme, as impressive as specimens of 200-year-old rockfish and 400-year-old clams may be, these long-lived species pale in comparison with deep-water corals that can live in excess of 4,000 years (Table 1.1). Imagine that when this coral first settled to begin life, Egyptian and Minoan cultures were flourishing, and Tutankhamun had not yet begun his reign. Indeed, this coral would have been over 1,000 years old when Buddha and Aristotle were born!

Some species are prolific and others are not. Many long-lived species produce few offspring after many years, in contrast to many species that produce millions of offspring, sometimes only months after they themselves

Table 1.1 Maximum age estimates of individuals from the wild in various marine taxa			
Taxa	Species	Location	Age (years)
Marine mammal	*Balaena mysticetus* (Bowhead whale)	Alaska, United States	211
Seabird	*Diomedea epomophora* (Royal albatross)	Unknown	>58
Marine reptile	*Chelonia mydas* (Green sea turtle)	Hawaii, United States	>>59
Marine fish	*Sebastes aleutianus* (Rougheye rockfish)	Unknown	205
Echinoderm	*Strongylocentrotus franciscanus* (Red sea urchin)	British Columbia, Canada	200
Bryozoan	*Melicerita obliqua*	Weddell Sea	45
Arthropod	*Homarus americanus* (American lobster)	Unknown	100
Pogonophoran	*Lamellibrachia luymesi*	Louisiana slope (~550 m)	250
Mollusk	*Arctica islandica* (Ocean quahog)	North Icelandic Shelf (80 m)	407
Cnidarian	*Leiopathes glaberrima* (Smooth black coral)	Oahu and Big Island, Hawaii, United States	4,265
Sponge	*Xestospongia muta* (Giant barrel sponge)	Curaçao	>2,300

were born. Mobility also varies. Corals are affixed to a single seafloor location, but other species move quickly and far. Sailfish can rocket through the water at 110 kilometers per hour. Atlantic bluefin tuna can move quickly, too, but take their transatlantic migrations of 5,800 kilometers at a more leisurely single kilometer per hour. Sooty shearwaters complete their 64,000 kilometer roundtrip migrations at 40 kilometers per hour, faster than most of us can move as we commute to work! The real marathon swimmers are humpback whales that complete 8,400-kilometer migrations, and Pacific tuna that make triple crossings of the Pacific.

Tricks of survival are many. Whereas species such as flounder can adjust their color to blend in with the environment and make themselves invisible to predators, nudibranch sea slugs alert predators that they are poisonous through bright colors. The diversity of life, its size, and its longevity reflect a wide range of adaptations evolved through time to survive in Planet Ocean (Figure 1.2).

Figure 1.2 The diversity of marine life.
These images from branches of the evolutionary tree of life illustrate the range of species and forms of life living in Planet Ocean.
(a) Giant sulfur bacteria inhabit anoxic sediments, devoid of oxygen, in the South Pacific. Though most bacteria require a microscope to be seen, these giants can be seen with the naked eye. See Chapter 8.
(b) *Halicreas minimum*, a jellyfish, drifting as zooplankton as deep as 300 meters. It lacks any stage that attaches to the seafloor. Its fertilized egg changes directly into a larval stage bearing tentacles, which then changes into the small medusa form in the photograph.
(c) The prized Atlantic bluefin tuna, *Thunnus thynnus,* swims in the open sea, hunting sardines, herring, mackerel, squid, and crustaceans. The Census tagged tuna to track their long migrations. See Chapter 7.

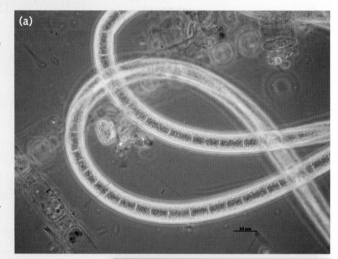
(a)

(b)

(c)

(d) Shown here is an amphipod, the small crustacean *Eusirus holmi*, that swims through the water as it searches for small prey to seize with its powerful, miniature, claws.

(e) In 2009 Census scientists found this flamingo tongue snail, *Cyphoma gibbosum*, near Grand Cayman, British West Indies. They listed it in the Gulf of Mexico biodiversity inventory amassed by the Census.

(f) The bioluminescent lure that anglerfish dangle in front of their mouths attracts potential prey and gives them their name. The ceratoid anglerfish of the genus *Lophodolos*, found by Census scientists, may be a new species.

Why a Census?

Beginning with environmentalist Rachel Carson's *The Sea Around Us*
and continued by Jacques Cousteau's films, interest in oceans grew
through the 1970s. Books and television brought the beauty of the ocean
into living rooms of homes around the world. The complexity of the *neritic*
(coastal) environment and the size and remoteness of the *pelagic* (offshore)
environment awed people. *Environment* is the sum total of physical, chemical,
and biotic factors (such as ocean currents, nutrients, prey) that act upon an
organism or an ecological community and ultimately determine its success
and failure. The environments that awed people supported fisheries that in
turn supported many economies around the world. But by the 1960s and
1970s, some fisheries were obviously declining, and as the environmental
movement took off, interest in living organisms was rising. Scientists began
to appreciate the array of life in the ocean, debated the patterns and causes
of diversity in coral reefs, and fueled interest in the myriad species that
had been discovered in the ocean.

Much of this interest was confined to a few specialists, and interest
in diversity and the underlying taxonomy declined by the early 1990s. Soon
after that, the rise of conservation biology, projections of millions of unknown
species in tropical rainforests, and E. O. Wilson's and J. F. (Fred) Grassle's
writings about diversity on land and sea revived interest in species diversity.

Wilson helped to bring to the public the term *biodiversity* for the
variability in genes, species, and ecosystems of a region, evolving it from
the more general term *diversity*. Although some scientists continued to study
biodiversity patterns in coastal areas and the deep sea, and the processes that
contributed to those patterns, major funding and effort to study marine
biodiversity did not materialize.

Meanwhile, the collapse of major fisheries once thought to be inexhaustible alarmed the public. Though explaining year-to-year variation in abundance had vexed biologists for a century, and although many fisheries had waxed
and waned over the years, the global scale and gravity of the fisheries collapse hit
home in the 1990s. The effects of removing the top predators from whales to
sharks to fishes cascaded onto other species. The cascade of effects from the
ocean surface to the seafloor provided evidence of unintended consequences
of fishing on ocean productivity. Effects on biodiversity extend far back in
time. These concerns shifted fisheries managers away from focusing on

single species to viewing entire ecosystems. They recognized the negative impacts of fishing on habitats and food webs.

An additional wrinkle was the new idea of ecosystem services. This concept refers to the key services from living organisms that benefit all life on Earth and, in many cases, benefit humans directly. In the ocean, these services include breakdown of sewage and other waste, nutrient cycling, shoreline stabilization, and provision of food and oils. Scientists argued that because biodiversity made ecosystems healthy, changes in biodiversity could diminish ecosystem services. This concern offered a practical application to move biodiversity research beyond a descriptive exercise to finding how oceans actually work, and then apply those findings to management. The role of biodiversity in ecosystem services matters because major changes have taken place and continue to occur in the ocean. Human production of pollutants and climate-change gases, plus fishing in the ocean caused many of the changes. Will changing biodiversity in the ocean influence ocean productivity and other ecosystem services?

Ocean biodiversity

During the late 1990s, some experts estimated that less than 5% of the biodiversity in the ocean had been described. Indeed, some estimated less than 1%. Even comparatively well-known seas, such as the shallow waters of northern Europe and the northeastern United States, continue to yield new discoveries of species and surprises about the life they contain. Much of the ocean remains unsampled. Biological data for the deep ocean beyond 200 meters depth exist for only a few square kilometers of the 300 million square kilometers of Planet Ocean. Much of the biodiversity on coral reefs remains to be sampled. And we are just now starting to learn about the diversity of microbes, whose unknown biodiversity could explode our estimates of the number of marine species.

During the last few decades, our understanding of life on Earth has broadened from one of life divided into five all-encompassing groupings to a new recognition of three domains. The first domain, the Eukarya or *eukaryotes*, encompasses the animals, plants, fungi, and microbial *protists*, (single-celled, simple organisms such as amoebas), all of which have nuclei and other specialized organelles within their cells. The other two domains, the Bacteria and bacteria-like, but different, Archaea, comprise the *prokaryotes* that lack most of these organelles. The broadened view of the diversity of life on Earth includes understanding how species have evolved. Molecular tools that tell how groups of organisms are related in an evolutionary context made this new view possible. Our view of life on Earth has been reorganized.

Counting all the fish in the sea

In 1995, the United States National Academy of Sciences emphasized the need to fill major gaps in understanding ocean life. The ensuing inaction discouraged Fred Grassle, a leading voice on marine biodiversity. One summer afternoon in 1996 he walked into the office of Jesse Ausubel, a program officer with the Alfred P. Sloan Foundation, and lamented the discouraging problem. They explored what to do about it. To some extent, a coincidence of geography – strong ties to Woods Hole Oceanographic Institution – brought Grassle and Ausubel to the same small town and marine science Mecca, where I wrote these chapters as a seasonal guest. As Rachel Carson noted almost 50 years ago, "Woods Hole is a wonderful place to come for research. Biologists come from all over. If you want to talk to them, you just come here in the summer instead of traveling all around the country to find them in winter."

At the time, the Sloan Foundation was supporting the first Digital Sky Survey to map the one hundred million objects in the sky. Grassle's concern intrigued Ausubel, and after further discussions, they developed the idea of "mapping" life in the ocean. Some weeks later, Ausubel strolled with one of his Sloan colleagues towards Aquinnah on the island of Martha's Vineyard and, while inhaling salt air and scanning the ocean edge, announced, "We've helped astronomers count all the stars in the sky; let's help marine scientists count all the fish in the sea." Appropriately, Aquinnah (formerly Gay Head) stood at the beginning of a sampling line established by Howard Sanders, one of the fathers of deep-sea biology, in the 1960s. The line ran all the way to Bermuda and its study changed our understanding of marine biodiversity. Grassle's passion for marine diversity, and especially the *invertebrates* or animals without backbones, ensured that "fish in the sea" would not be taken literally, and his dream of a global program in marine biodiversity began to become a reality.

During the next three years, marine experts from around the world gathered to talk, identify research gaps, and formulate a strategy to understand life in the ocean. They came from wealthy countries and poor ones, from polar research labs and the tropics, and with interests from whales to bacteria. They met in marine centers around the United States and United Kingdom, and in Greece and Thailand. A plan emerged to undertake an unprecedented 10-year census of the world's marine life in an international program called the Census of Marine Life. The overarching goal of the Census would be to understand the diversity, distribution, and abundance of marine organisms across all ocean realms from the shoreline to the ocean abyss, and to consider their past, present, and future. The task was formidable, because the concept was global in scope and aimed to sample the vast oceans and their diversity of life. Unlike the censuses that count the single species of humans with fixed addresses in single nations, this Census would consider all species and ocean habitats. Most of the ocean has never been sampled or seen because much of it is thousands of kilometers from land and several kilometers deep. And some of the most species-rich waters of the ocean are near developing countries with little funding and infrastructure for research and exploration.

The ocean constantly changes as the sun rises and sets, as wind blows and seasons change, and as such phenomena as El Niño wax and wane. Many waters straddle national and international boundaries and organisms move through entire ocean basins in a matter of weeks and months. Until the Census began, marine biology lacked coordination in biodiversity research, particularly internationally. Scientists had worked together on fisheries problems in groups such as the International Council for the Exploration of the Sea, but the groups focused largely on single species targeted by fisheries. Uncoordinated efforts to study biodiversity produced findings, but could not capture the broad variability in space and time of marine diversity, distribution, and abundance in the global ocean as the planned Census hoped to do.

Bringing together experts from around the world facilitated coordination. The gatherings introduced individuals from developing nations with limited equipment, ships, and knowledge of sampling, to experts from developed countries with more scientists, resources, and links to multinational programs. The gathered scientists discussed objectives, oceanographic cruises, sampling, and analytical methodologies so they could work together to tackle big questions that no individual or nation could hope to answer alone.

Indeed, the crowning achievement of the Census may not be the thousands of scientific papers it has catalyzed and their many findings. Rather the crown may be exciting and unifying global researchers toward the common objective of understanding life in the ocean and managing ocean resources effectively.

Although programs such as the Human Genome Project or the World Ocean Circulation Experiment (WOCE) have tackled big questions through international collaboration, they have typically addressed a single big question with one set of techniques. The Census has instead used tools from rubber boots to robots and molecular biology to satellites to study the diversity, distribution, and abundance of microbes to whales. From fishermen concerned about declining catches to conservationists worried about extinctions and habitat loss, all these groups stand to benefit from coordinated studies of the oceans and communication of scientific findings. The legacy of cooperation and collaboration represents a new way of doing integrative science at a truly global scale that will live long after the formal Census ends. From the seafood aisle at the supermarket to tourists snorkeling the reefs outside their beachfront hotel, the legacy will provide enduring benefits.

The Census of Marine Life

The complexity and scale of the Census attracted thousands of experts from more than 80 nations (Figure 1.3), who rallied around understanding marine biodiversity, distribution, and abundance in the past, present, and future. They organized into 17 interlinked projects (Table 1.2 and summarized on the inside back cover) that divided up the scientific tasks to focus on contrasting ocean habitats, groups of organisms, regions, and how the ocean has changed in the past and will change in the future (Figure 1.4). They then worked together to synthesize the many findings into an understanding of the global ocean. In the largest study ever undertaken of marine life, this team developed new technologies and partnerships to generate knowledge that encompassed new species and explored habitats and the movements and patterns of biota. It evaluated how abundance in the ocean is changing and might continue to change. In addition to the projects under the umbrella of the Census, this excitement has also benefited other collaborative and complementary programs to study marine biodiversity and ocean life.

Several thousand new species and growing – some formally named and some not – were discovered during the 10-year life of the Census, including the beautiful and bizarre, discovered in many parts of the ocean and from organisms spanning microbes to fishes (Figure 1.5). Indeed, the increase in the number of marine species may accelerate, as 5–10 years often intervene between when a specimen is collected and when it is formally described and honored with a name. Studies have expanded the knowledge of the diversity of marine species and the evolutionary relationships among them. We know more about where they live and why they live there. The public has joined in the excitement, fascinated by wondrous new species, voicing concern over declines in species that they feed their children, and embracing the beauty of the ocean. We also now see more clearly what we know is unknown and what is "unknowable" with current technology and effort. Understanding the limits to knowledge of marine biodiversity guides science to what we can and must do. The Census has located and can now return to ocean environments, and their "hot spots" and "cold spots" of elevated or reduced biodiversity and abundance. We know where to look, and what ocean environments or habitats will continue to yield discoveries.

Figure 1.3 The reach of the global Census of Marine Life.
The map highlights in yellow more than 80 countries that participated in the Census, encompassing most large coastal nations. Coastal nations in white have limited marine scientific capacity and their coastal waters remain largely unknown, even after the 10-year global effort by the Census.

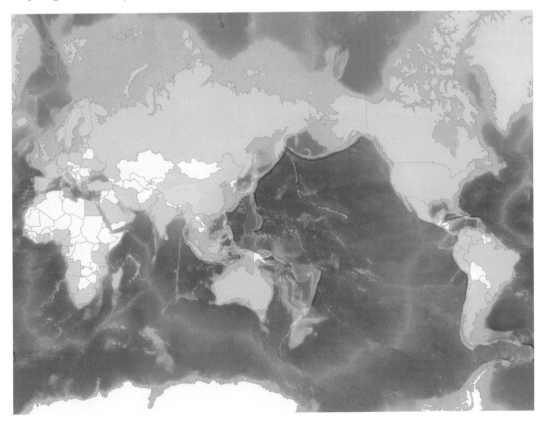

Table 1.2 The projects within the Census of Marine Life

Project Name	Objective
Oceans Past	
History of Marine Animal Populations (HMAP)	Use historical archives to analyze marine population data before and after significant human impacts on the ocean.
Oceans Present: Geographic Realms	
Natural Geography in Shore Areas (NaGISA)	Inventory and monitor biodiversity in the narrow nearshore zone of the world's oceans at depths of less than 20 meters.
Gulf of Maine Area (GoMA)	Document all species in a regional ecosystem and show how understanding biodiversity patterns and factors improves ecosystem management.
Census of Coral Reef Ecosystems (CReefs)	Collaborate internationally for a global census of coral reefs.
Continental Margin Ecosystems on a Worldwide Scale (COMARGE)	Assemble and understand the patterns of diversity on continental margins (slopes).
Patterns and Processes of the Ecosystems of the Northern Mid-Atlantic (MAR-ECO)	Study the animals and their distribution on the northern mid-Atlantic Ridge.
Global Census of Marine Life on Seamounts (CenSeam)	Explore seamount ecosystems, globally.
Census of Diversity of Abyssal Marine Life (CeDAMar)	Document diversity patterns on deep-sea abyssal plains.
Biogeography of Deep-Water Chemosynthetic Ecosystems (ChEss)	Study the biogeography of deep-water chemosynthetic ecosystems.
Arctic Ocean Diversity (ArcOD)	Inventory biodiversity in the Arctic sea ice, water column, and seafloor from shallow shelves to deep basins.
Census of Antarctic Marine Life (CAML)	Survey life in the cold Southern Ocean surrounding Antarctica.
Oceans Present – Global Distributions	
International Census of Marine Microbes (ICoMM)	Document microbial diversity and build a cyberinfrastructure to index and organize what is known about microbes, which account for up to 90% of ocean biomass.
Census of Marine Zooplankton (CMarZ)	Assess biodiversity of the animal plankton, including ~7,000 described species representing 15 different phyla.
Oceans Present – Animal Movements	
Pacific Ocean Shelf Tracking Project (POST)	Develop and apply new electronic tags on salmon and other species to study how they migrate along the Pacific coastline.
Tagging of Pacific Predators (TOPP)	Apply new electronic tagging to track the migration of large, open-ocean animals.

Table 1.2 (*cont.*)	
Project Name	Objective
Oceans Future	
Future of Marine Animal Populations (FMAP)	Project globally changing abundance, distribution, and diversity and the effects of fishing, climate change, and other influences.
Using the Data	
Ocean Biogeographic Information System (OBIS)	Build an Internet-based bank of spatial information on marine species that brings together datasets from all around the world, adds new data, and is open to all to visualize how species relate and react to their environment.

Figure 1.4 Seventeen Census projects spanning Planet Ocean.
The map locates where Census projects studied marine life, from near shore to mid-ocean, from the Arctic to Antarctica, and from the shallow waters to the deepest ocean.

CENSUS OF MARINE LIFE PROJECT AREAS

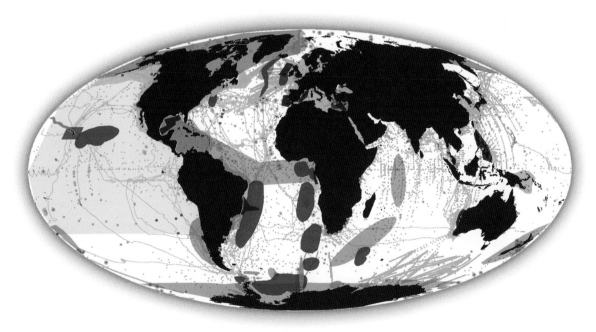

Coastal	Polar	Pelagic	Deep Sea	Global Information and Anaylsis
Regional Ecosystems (GoMA)	Arctic Ocean (ArcOD)	Top Predators (TOPP)	Vents and Seeps (ChEss)	Oceans Future (FMAP)
Near Shore (NaGISA)	Antarctic Ocean (CAML)	Continental Shelves (POST)	Abyssal Plains (CeDAMar)	Information System (OBIS)
Coral Reefs (CReefs)		Zooplankton (CMarZ)	Seamounts (CenSeam)	
		Microbes (ICoMM)	Continental Margins (COMARGE)	Oceans Past (HMAP)
			Mid-Ocean Ridges (MAR-ECO)	

Figure 1.5 Spectacular new species.

Examples of the thousands of new species discovered by the Census, spanning tiny crustaceans the size of a pinhead to large kelp and lobster, from the shoreline to the abyssal plain.

(a) South of Easter Island, Census explorers at hydrothermal vents discovered a crab so unusual it warranted establishing a whole new family, Kiwaidae, for a new genus, *Kiwa*. Census scientists named the family and genus for the mythological Polynesian goddess of shellfish. Its furry or hairy appearance inspired the *hirsuta* in the second half of its species name, and also its common name, Yeti crab.

(b) The small, mostly transparent ghost shrimp, *Vulcanocalliax arutyunovi*, is a new species from the Captain Arutyunov mud volcano in the Gulf of Cadiz. The specimen pictured is an egg-bearing female. See Chapter 8.

(c) Even in shallow water along Alaska's Aleutian Islands, Census researchers discovered new species, such as this kelp, *Aureophycus aleuticus*, which grows 3 meters long.

(d) This blind lobster belongs to the new genus *Dinochelus*, which means "mighty claws." Because systematists have the privilege of choosing the name of the new species they describe, they named it *Dinochelus ausubeli*, in honor of Census co-founder Jesse Ausubel "in recognition of his vision and support for marine biodiversity exploration." Shane Ahyong, Tin-Yam Chan, and Philippe Bouchet named the species, which was discovered at 300 meters depth during a Census expedition to the Philippine Sea.

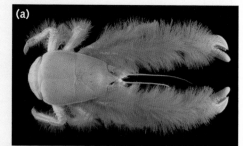

(a)

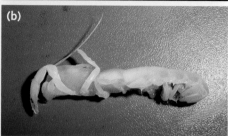

(b)

(d)

(c)

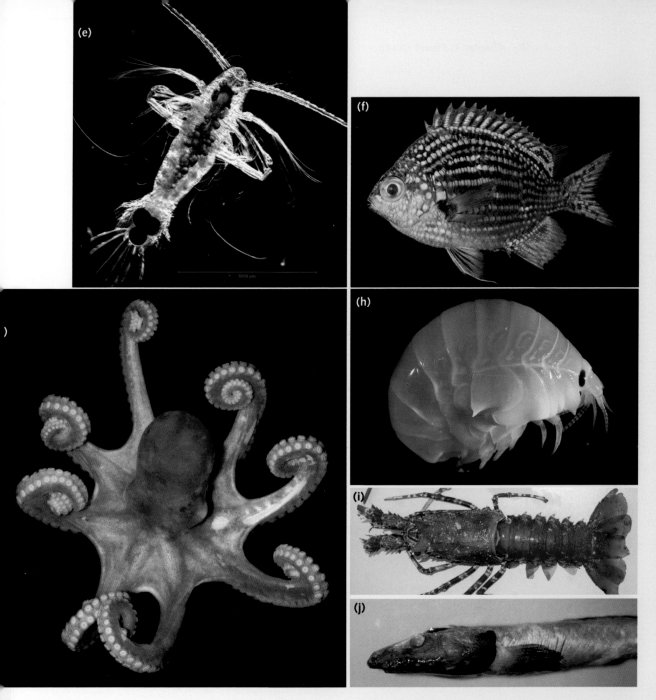

(e) Although the Arctic deep-water copepod crustacean, *Euaugaptilus hyperboreus*, swims weakly, it is classified as plankton, drifting with the current.

(f) Scientists found the deep blue chromis in 1997 in deep Pacific reefs near Palau. Although its habitat below 100 meters may be short of abyssal depth, they called it *Chromis abyssus* because it lives deeper than most damselfish. *Chromis* describes its striking color.

(g) Named by Census scientists for their late colleague Robin Rigby, the octopod *Benthoctopus rigbyae* grows up to 40 centimeters long and lives at depths of 250–600 meters, near the Antarctic Peninsula.

(h) Census scientists placed the new species of amphipod or water flea in the genus *Shackletonia*, which honors Antarctic explorer Ernest Shackleton.

(i) Lobster fishermen surprised Census scientists in South Africa. The fishermen brought in a spiny lobster caught on Walters Shoals along the Madagascar Ridge, which scientists realized was a new species *Palinurus barbarae*. See Chapter 3.

(j) *Lycodonus* sp., a new species of eelpout collected near the Mid-Atlantic Ridge, illustrates that even the familiar fishes continue to yield new species.

BIBLIOGRAPHY

Abele, L. G. and K. Walters, Marine benthic diversity: critique and alternative explanation. *J. Biogeogr.*, **6**:2 (1979), 115–26.

Amaral-Zettler, L., L. F. Artigas, J. Baross, *et al.*, A global census of marine microbes. In A. D. McIntyre (ed.), *Life in the World's Oceans: Diversity, Distribution, and Abundance* (Oxford: Blackwell Publishing Ltd., 2010), pp. 223–45.

Angel, M. V., What is the deep sea? In D. J. Randall and A. P. Farrell (eds.), *Deep-sea Fishes* (San Diego: Academic Press, 1997), pp. 1–41.

Auster, P. J., R. J. Malatesta, R. W. Langton, *et al.*, The impacts of mobile fishing gear on seafloor habitats in the Gulf of Maine (Northwest Atlantic): implications for conservation of fish populations. *Rev. Fish Sci.*, **4**:2 (1996), 185–202.

Bader, B. and P. Schafer, Skeletal morphogenesis and growth check lines in the Antarctic bryozoan *Melicerita obliqua. J. Nat. Hist.*, **38**:22 (2004), 2901–22.

Balazs, G. H., *Synopsis of Biological Data on the Green Turtle in the Hawaiian Islands* (Honolulu, Hawaii: NMFS, 1980).

Bergquist, D. C., F. M. Williams and C. R. Fisher, Longevity record for deep-sea invertebrate: the growth rate of a marine tubeworm is tailored to different environments. *Nature*, **403**:6769 (2000), 499–500.

Block, B. A., D. P. Costa, and S. J. Bograd, A view of the ocean from Pacific predators. In A. D. McIntyre (ed.), *Life in the World's Oceans: Diversity, Distribution, and Abundance* (Oxford: Blackwell Publishing Ltd., 2010, pp. 291–311.

Botsford, L. W., J. C. Castilla, and C. H. Peterson, The management of fisheries and marine ecosystems. *Science*, **277**:5325 (1997), 509–15.

Butman, C. A., J. T. Carlton, and S. R. Palumbi, Whaling effects on deep-sea biodiversity. *Conserv. Biol.*, **9**:2 (1995), 462–4.

Cailliet, G. M., A. H. Andrews, E. J. Burton, *et al.*, Age determination and validation studies of marine fishes: do deep-dwellers live longer? *Exp Gerontol*, **36**:4–6 (2001), 739–64.

Carson, R., *The Sea Around Us* (New York: Oxford University Press, 1951).

Carson, R., *Silent Spring* (Boston: Houghton Mifflin, 1962).

Connell, J. H., Diversity in tropical rain forests and coral reefs: high diversity of trees and corals is maintained only in a non-equilibrium state. *Science*, **199**:4335 (1978), 1302–10.

Costanza, R., The ecological, economic, and social importance of the oceans. *Ecol. Econ.*, **31**:2 (1999), 199–213.

Daily, G. C., *Nature's Services: Societal Dependence on Natural Ecosystems* (Washington, DC: Island Press, 1997).

Dayton, P. K. and R. R. Hessler, Role of biological disturbance in maintaining diversity in deep sea. *Deep Sea Res.*, **19**:3 (1972), 199.

Diaz, J. M., F. Gast and D. C. Torres, Rediscovery of a Caribbean living fossil: *Pholadomya candida* GB Sowerby I, 1823 (Bivalvia: Anomalodesmata: Pholadomyoidea). *Nautilus*, **123**:1 (2009), 19–20.

Ebbe, B., D. Billett, A. Brandt, *et al.*, Diversity of abyssal marine life. In A. D. McIntyre (ed.), *Life in the World's Oceans: Diversity, Distribution, and Abundance* (Oxford: Blackwell Publishing Ltd., 2010), pp. 139–60.

Ebert, T. A. and J. R. Southon, Red sea urchins (*Strongylocentrotus franciscanus*) can live over 100 years: confirmation with A-bomb ^{14}carbon. *Fish Bull.*, **101**:4 (2003), 915–22.

Erwin, T. L., Tropical forests: their richness in Coleoptera and other arthropod species. *Coleopterists Bull.*, **36** (1982), 74–5.

Etter, R. J. and J. F. Grassle, Patterns of species diversity in the deep sea as a function of sediment particle size diversity. *Nature*, **360**:6404 (1992), 576–8.

Finch, C. E., *Longevity, Senescence, and the Genome* (Chicago: University of Chicago Press, 1990).

Frank, K. T., B. Petrie, J. S. Choi, and W. C. Leggett, Trophic cascades in a formerly cod-dominated ecosystem. *Science*, **308**:5728 (2005), 1621–3.

George, J. C., J. Bada, J. Zeh, *et al.*, Age and growth estimates of bowhead whales (*Balaena mysticetus*) via aspartic acid racemization. *Can. J. Zool.*, **77**:4 (1999), 571–80.

Grassle, J. F., *Species Diversity, Genetic Variability and Environmental Uncertainty*. Fifth European Marine Biology Symposium, Piccin, Padua (1972).

Grassle, J. F., A plethora of unexpected life. *Oceanus*, **31**:4 (1988), 41–6.

Grassle, J. F., Species-diversity in deep-sea communities. *Trends Ecol. Evol.*, **4**:1 (1989), 12–5.

Grassle, J. F., Deep-sea benthic biodiversity. *Bioscience*, **41**:7 (1991), 464–9.

Grassle, J. F. and H. L. Sanders, Life histories and role of disturbance. *Deep Sea Res.*, **20**:7 (1973), 643–59.

Hjort, J., Fluctuations in the great fisheries of Northern Europe. *Rapports et Proces-Verbaux des Reunions du Conseil International pour l'Exploration de la Mer*, **20** (1914), 1–228.

Holm, P., History of marine animal populations: a global research program of the Census of Marine Life. *Oceanol. Acta*, **25**:5 (2003), 207–11.

del Hoyo, J., A. Elliot, and J. Sargatal (eds.), *Handbook of the Birds of the World – Ostrich to Duck* (Barcelona: Lynx Edicions, 1992).

Huston, M., General hypothesis of species-diversity. *Am. Nat.*, **113**:1 (1979), 81–101.

Knowlton, N., Coral reef coda: what can we hope for? In I. Cote and J. Reynolds (eds.), *Coral Reef Conservation* (Cambridge, UK: Cambridge University Press, 2006), pp. 538–49.

Lambshead, P. J. D., Recent developments in marine benthic biodiversity research. *Oceanis*, **19**:6 (1993), 5–24.

Langton, R. W. and P. J. Auster, Marine fishery and habitat interactions: to what extent are fisheries and habitat interdependent? *Fisheries*, **24**:6 (1999), 14–21.

Levin, L. A., C. L. Huggett, and K. F. Wishner, Control of deep-sea benthic community structure by oxygen and organic-matter gradients in the Eastern Pacific Ocean. *J. Mar. Res.*, **49**:4 (1991), 763–800.

Loreau, M., S. Naeem, P. Inchausti, *et al.*, Ecology: biodiversity and ecosystem functioning: current knowledge and future challenges. *Science*, **294**:5543 (2001), 804–8.

Lotze, H. K., H. S. Lenihan, B. J. Bourque, *et al.*, Depletion, degradation, and recovery potential of estuaries and coastal seas. *Science*, **312**:5781 (2006), 1806–9.

MBL, Rachel L. Carson: *Centennial 1907–2007*. Woods Hole, MA: MBL: Biological Diversity in Woods Hole; 2010 (cited 2010). Available from: http://www.mbl.edu/news/features/pdf/carson_article_scanlon.pdf.

McMurray, S. E., J. E. Blum, and J. R. Pawlik, Redwood of the reef: growth and age of the giant barrel sponge *Xestospongia muta* in the Florida Keys. *Mar. Biol.*, **155**:2 (2008), 159–71.

Miloslavich, P., The Census of Marine Life in the Caribbean: A Biodiversity Program. *Rev. Biol. Trop.*, **56** (2008), 171–81.

Musick, J. A., Criteria to define extinction risk in marine fishes: the American Fisheries Society initiative. *Fisheries*, **24**:12 (1999), 6–14.

Myers, R. A., J. K. Baum, T. D. Shepherd, S. P. Powers, and C. H. Peterson, Cascading effects of the loss of apex predatory sharks from a coastal ocean. *Science*, **315**:5820 (2007), 1846–50.

Naeem, S., J. M. H. Knops, D. Tilman, *et al.*, Plant diversity increases resistance to invasion in the absence of covarying extrinsic factors. *Oikos*, **91**:1 (2000), 97–108.

National Research Council, *Understanding Marine Biodiversity: A Research Agenda for the Nation* (Washington, DC: National Academy Press, 1995).

Pages, F., P. Flood, and M. Youngbluth, Gelatinous zooplankton net-collected in the Gulf of Maine and adjacent submarine canyons: new species, new family (Jeanbouilloniidae), taxonomic remarks and some parasites. *Sci. Mar.*, **70**:3 (2006), 363–79.

Paterson, G. L. J., Patterns of polychaete assemblage structure from bathymetric transects in the Rockall Trough, NE Atlantic Ocean. PhD thesis, University of Wales (1993).

Pauly, D., V. Christensen, J. Dalsgaard, R. Froese, and F. Torres, Fishing down marine food webs. *Science*, **279**:5352 (1998), 860–3.

Poore, G. C. B. and G. D. F. Wilson, Marine species richness. *Nature*, **361**:6413 (1993), 597–8.

Quijon, P. A. and P. V. R. Snelgrove, Predation regulation of sedimentary faunal structure: potential effects of a fishery-induced switch in predators in a Newfoundland sub-Arctic fjord. *Oecologia*, **144**:1 (2005), 125–36.

Ramirez-Llodra, E., Fecundity and life-history strategies in marine invertebrates. *Adv. Mar. Biol.*, **43** (2002), 88–170.

Rasmussen, K., D. M. Palacios, J. Calambokidis, *et al.*, Southern Hemisphere humpback whales wintering off Central America: insights from water temperature into the longest mammalian migration. *Biol. Lett.*, **3**:3 (2007), 302–5.

Reaka-Kudla, M. L., The global biodiversity of coral reefs: a comparison with rain forests. In M. L. Reaka-Kudla, D. E. Wilson, and E. O. Wilson (eds.), *Biodiversity II: Understanding and Protecting Our Biological Resources* (Washington, DC: Joseph Henry Press, 1997), pp. 83–108.

Reaka-Kudla, M. L., Biodiversity of Caribbean coral reefs. In P. Miloslavich and E. Klein (eds.), *Caribbean Marine Biodiversity: The Known and the Unknown* (Lancaster, PA: DEStech Publications, 2005), pp. 259–76.

Rex, M. A., C. T. Stuart, R. R. Hessler, *et al.*, Global-scale latitudinal patterns of species diversity in the deep-sea benthos. *Nature*, **365**:6447 (1993), 636–9.

Rice, J. C. and M. J. Rochet, A framework for selecting a suite of indicators for fisheries management. *ICES J. Mar. Sci.*, **62**:3 (2005), 516–27.

Roark, E. B., T. P. Guilderson, R. B. Dunbar, S. J. Fallon, and D. A. Mucciarone, Extreme longevity in proteinaceous deep-sea corals. *Proc. Natl. Acad. Sci. USA*, **106**:13 (2009), 5204–8.

Rooker, J. R., J. R. A. Bremer, B. A. Block, *et al.*, Life history and stock structure of Atlantic bluefin tuna (*Thunnus thynnus*). *Rev. Fish Sci.*, **15**:4 (2007), 265–310.

Roy, K., D. Jablonski, J. W. Valentine, and G. Rosenberg, Marine latitudinal diversity gradients: tests of causal hypotheses. *Proc. Natl. Acad. Sci. USA*, **95**:7 (1998), 3699–702.

Sanders, H. L., Marine benthic diversity: a comparative study. *Am. Nat.*, **102**:925 (1968), 243.

Sanders, H. L., Benthic marine diversity and stability-time hypothesis. *Brookhaven Symp. Biol.*, **22** (1969), 71.

Shaffer, S. A., Y. Tremblay, H. Weimerskirch, *et al.*, Migratory shearwaters integrate oceanic resources across the Pacific Ocean in an endless summer. *Proc. Natl. Acad. Sci. USA*, **103**:34 (2006), 12799–802.

Smith, C. R., D. J. Hoover, S. E. Doan, *et al.*, Phytodetritus at the abyssal seafloor across 10° of latitude in the central equatorial Pacific. *Deep Sea Res. II*, **43**:4–6 (1996), 1309–38.

Snelgrove, P., Why care about marine biodiversity? *Sea Technol.*, **37**:9 (1996), 93.

Snelgrove, P., T. H. Blackburn, P. A. Hutchings, *et al.*, The importance of marine sediment biodiversity in ecosystem processes. *Ambio*, **26**:8 (1997), 578–83.

Snelgrove, P. V. R., Getting to the bottom of marine biodiversity: sedimentary habitats. *BioScience*, **49** (1999), 129–38.

Snelgrove, P. V. R., J. F. Grassle, and R. F. Petrecca, The role of food patches in maintaining high deep-sea diversity: field experiments with hydrodynamically unbiased colonization trays. *Limnol. Oceanogr.*, **37**:7 (1992), 1543–50.

Snelgrove, P. V. R., J. F. Grassle, and R. F. Petrecca, Experimental evidence for aging food patches as a factor contributing to high deep-sea macrofaunal diversity. *Limnol. Oceanogr.*, **41**:4 (1996), 605–14.

Sogin, M. L., H. G. Morrison, J. A. Huber, *et al.*, Microbial diversity in the deep sea and the underexplored "rare biosphere." *Proc. Natl. Acad. Sci. USA*, **103** (2006), 12115–20.

Soule, M. E. and K. Kohm, *Research Priorities for Conservation Biology* (Washington, DC: Island Press, 1989).

Soule, M. E. and B. A. Wilcox (eds.), *Conservation Biology: An Evolutionary-Ecological Perspective* (Sunderland, MA: Sinauer Associates, 1980).

Taggart, C., T. J. Anderson, C. Bishop, *et al.*, Overview of cod stocks, biology, and environment in the Northwest Atlantic region of Newfoundland, with emphasis on northern cod. *ICES J. Mar. Sci. Symp.*, **198** (1994), 140–57.

Thistle, D. and J. E. Eckman, The effect of biologically produced structure on the benthic copepods of a deep-sea site. *Deep Sea Res. A*, **37**:4 (1990), 541–54.

Tilman, D., D. Wedin, and J. Knops, Productivity and sustainability influenced by biodiversity in grassland ecosystems. *Nature*, **379**:6567 (1996), 718–20.

Van Valen, L., Energy and evolution. *Evol. Theory*, **1** (1976), 179–229.

Vermeulen, N., *Supersizing Science: On Building Large-scale Research Projects in Biology* (Maastricht: Maastricht University Press, 2009).

Wanamaker, A. D., J. Heinemeier, J. D. Scourse, *et al.*, Very long-lived mollusks confirm 17th century AD tephra-based radiocarbon reservoir ages for North Icelandic shelf waters. *Radiocarbon*, **50**:3 (2008), 399–412.

Wilson, E. O., *The Diversity of Life* (Cambridge, MA: Harvard University Press, 1992).

Worm, B., H. K. Lotze, and R. A. Myers, Ecosystem effects of fishing and whaling in the North Pacific and Atlantic Ocean. In J. A. Estes, R. L. Brownell, D. P. DeMaster, D. F. Doak, and T. M. Williams (eds.), *Whales, Whaling, and Ocean Ecosystems* (Berkeley, CA: University of California Press, 2006).

Worm, B. and R. A. Myers, Meta-analysis of cod-shrimp interactions reveals top-down control in oceanic food webs. *Ecology*, **84**:1 (2003), 162–73.

2

The ocean environments

The diversity of habitats in Planet Ocean has driven the evolution of life, its size, life span, mobility, and survival. We think of salt water and its immensity, perhaps seen from an airplane on an intercontinental flight, few seeing the distinct environments beneath the waves. When we think of what lives beneath the waves, we might think of whales and sharks rather than the 90% of total ocean biomass that is microbes.

A group of marine ecologists organized all the five oceans of the world (Figure 1.1) and their habitats into 232 *ecoregions*, which have relatively homogeneous species composition, clearly distinct from adjacent ecoregions. Variables such as surface currents and bottom flow, which in turn define a suite of key variables from seafloor sediments to water temperature, differentiate these ecoregions. All form "One Ocean" that covers "Planet Ocean" and is "Home Sweet Home" to species evolving to adapt to their distinct habitats.

Scientists divide the ocean into the *near shore* close to land and the *continental shelf* less than about 200 meters deep. Further offshore, the open ocean sits above the continental slope 200–4,000 meters beneath that flattens to the abyssal plains at 4,000–6,000 meters below the surface. Along with deep-ocean trenches as deep as 11,000 meters, the slope and abyssal plains comprise the deep sea. Some argue that the true deep sea, where the ocean's surface and bottom are almost disconnected and the influence of land typically fades, occurs near 1,000 meters depth, but the precise demarcation varies with location and whom you ask. For those rolling around on storm-battered ships hundreds of kilometers from shore, this distinction is moot.

The reality is that ecoregions form a continuum that encompasses a variety of habitats, each with its own characteristics and, in many cases, suites of species.

Sampling the ocean

Some parts of the ocean are relatively easy to sample. Because the shoreline between high and low tides is easily accessible without ships or complex equipment, even aboriginal man studied these habitats. Scuba divers can safely access the upper 20 meters, but depths much greater than that require specialized and expensive sampling gear, including boats, ships, and oceanographic instruments. The development of satellite sensors beginning in the late 1970s provided broad views of the single-celled, plant-like photo-synthetic organisms (*phytoplankton*) that live in the surface waters, and new views of the temperature and circulation over large swaths of ocean. These views spanned ocean basins from the near shore to the open ocean.

But while satellites see wide swaths, they only see the upper few meters or tens of meters of the ocean, depending on the specific sensor used. Satellites cannot see most of the ocean because it is some 3,800 meters deep on average. Thus, most organisms are invisible to satellites. For example, they largely miss the *zooplankton* ("drifting animals") that include a rich array of invertebrate jellyfish, shrimp-like crustaceans, larval fishes, and other organisms. Though most are only centimeters or less in size, many are abundant. Though the shallowest seafloor can be seen with some types of special equipment mounted on satellites and airplanes, water hides most of the seafloor and the life in and above it.

Traditional nets, bottles, and cameras that are lowered over the side of the ship (see Chapter 4) have sampled the organisms that live farther from shore in the opaque depths that comprise the more than 99% of the ocean that satellites cannot see. Wind, waves, and surface currents push the ship around as the samples are collected, adding to the challenge. It's like trying to sample birds and worms on land from a hot air balloon on a cloudy, windy night.

The last decade has seen advances in ocean sampling as technologies such as satellites, submersibles, and autonomous underwater vehicles (AUVs) extended traditional sampling from ships. Submersibles and remotely operated, unmanned vehicles (ROVs) lowered from a ship and monitored from the surface open new "viewing ports" for sampling organisms and measuring environmental conditions. These vehicles are expensive, however, and few are available for scientific research. Even the oceanographic mother ships themselves are expensive and few. Because funds and distance have limited past research, sampling has often been during brief cruises, bringing back just snapshots and anecdotes about a tiny fraction of the ocean through space and in time.

Now, marine animals themselves carry sensors, "listening devices" enumerate organisms as they swim by, and other acoustic devices scan wide swaths of ocean water and seafloor (see Chapters 7 and 8). New imaging and statistical tools provide fresh views of data, and new molecular barcoding rapidly identifies species. Amid the staggering array of life in Planet Ocean in 2010, new tools are moving some key unknowns into the known, and some of the unknowable into the knowable.

Near the land

The nearshore and coastal zone, defined here as the seabed and the water above it from the upper intertidal zone to the edge of the continental shelf (between 100 and 200 meters deep, depending on location), covers only about 7% of Earth's seafloor. Nevertheless, together they support important habitats, including many fisheries such as Georges Bank and tourist destinations such as the Caribbean and Thailand, as well as waters rich and poor in biodiversity. Coral reefs, for example, support more species per unit area than any other marine habitat and potentially any habitat on Earth (Figure 2.1), whereas intertidal environments that teem with abundance have relatively few species and little diversity.

The rocky intertidal habitats at all latitudes support species specifically adapted to those environments. Terrestrial approaches to understanding biodiversity are now applied in intertidal ecosystems. Because of their accessibility, these environments are ideal laboratories to study how climate change, invasive species, shoreline development, and pollution are changing patterns of biodiversity and the ramifications of these changes. Natural fluctuations in temperature and salinity, strong waves, sediment scouring (and sometimes ice) disturbance, and exposure to air in the shallow nearshore and coastal waters create harsher environments than those in deeper water. Many *benthic* species, those that are adapted to life on the seafloor, cannot tolerate the fluctuations. Species that can tolerate these conditions, however, find abundant light for photosynthesis and fixed substrate for *sessile*

Figure 2.1 Coral reefs, the rainforests of the sea.
A large Napoleon wrasse chasing a school of small anthias reef fish swirling above the Elphinstone reef in the Red Sea illustrates the diversity of life in the tropical coastal oceans. The complex structure of coral at the bottom of the image shelters a riot of hidden species. See Chapter 5. This image was shot for the Galatée film *Oceans*. See Chapter 4.

(attached, non-mobile) organisms to settle on and grow. These habitats yield much *biomass* (total mass of living organisms) of *benthos* (bottom-living organisms) that includes seaweeds, seagrasses and other plants, and invertebrates such as barnacles and mussels. In addition to the abundance of food, the hiding places around these fixed organisms provide protective habitat for a range of species that live above the seafloor – especially the young that are vulnerable to predators.

Estuaries have played a grand role in human history because they are where rivers meet the ocean, and the junction may provide drinking water and remove waste. At the junction, seaports load people, raw materials, and goods from oil to autos. Almost two-thirds of the largest cities in the world are built around estuaries (think New York, London, Athens, Tokyo, Vancouver). Many organisms that live in the extremes of freshwater or fully saline saltwater cannot tolerate the variable and brackish water in estuaries. The tolerant species, however, may grow abundantly on nutrients eroded from the land. Seagrasses, salt marshes, and mangroves shelter a variety of species, including vulnerable young. Not surprisingly, excess nutrients and hypoxia, pollutants, overfishing, altered hydrography, and invasive species affect estuaries. Although in 1664 alewife ascended the rivers of New England "in such multitudes...as will scarcely allow them to swim," most Boston residents know "Alewife" best as the name of a subway stop.

Nearshore ecoregions buried in mud and sand are less well studied than the rocky intertidal, but also matter for organisms and their productivity and nutrient cycling. Although buried habitats are also threatened by human activities, the cover of sand and mud teems with small invertebrates such as polychaete worms, crustaceans, and mollusks. This buried life feeds migratory birds and groundfish such as cod and halibut.

In mangrove forests in the tropics and salt marshes in middle latitudes, plants bind mud and create productive habitats that are home to specialized species, often as a nursery. Although many species in mangroves are marine, the vegetation that rises above the ocean's surface also supports insects and birds, some of which feed on the small invertebrates in the sediment. Humans gather at the coastline; 3.6 billion people already live within 150 kilometers of the ocean, putting the near shore at risk. The gathering multitude intrudes into mangroves and salt marshes on the ocean edge, clearing plants for coastal development. Because shrimp *mariculture*, aquaculture in the ocean, displaces mangroves, it lessens biodiversity.

The shallow subtidal receives sufficient sunlight for photosynthesis by both single-celled phytoplankton and multicellular seagrasses, kelps, or other seaweeds attached to the seafloor. People who spend time on the shoreline know partially submerged seagrass meadows along sheltered shorelines are often sensitive and important nurseries for juvenile fishes, feeding seabirds, and other species. Kelps, the large brown seaweeds with broad blades akin to leaves, support fishes and small animals that graze on them, as well as sea otters and other species that eat the smaller animals (Figure 2.2). Some subtidal habitats are less known, such as the hard-bottom *rhodolith beds* created by calcareous red algae that encrust the seafloor and, in turn, provide habitat for other species. All nearshore habitats support life

adapted to that environment, and although some species live in a range
of habitat types, others are specific in their needs and thus more vulnerable
when habitat is lost.

Rich fisheries abound in nearshore areas. Hunting and fishing in the
Wadden Sea for birds, marine mammals, fishes, and mollusks have fed
coastal communities. Herring fisheries in northwest Europe, cod fisheries
in the eastern United States, and salmon fisheries of the White and Barents
Sea, that shaped nations, have dramatically declined, creating hardship.

I witnessed one of the "textbook" collapses in my own backyard. Explorers
in the late fifteenth century caught Newfoundland cod just by lowering baskets
over the side of the ship. But by 1992 fish were so scarce that fishing closed,
as fishermen, scientists, and managers unanimously agreed that the stocks
had collapsed. Overnight, the closure threw fully 22,000 of Newfoundland's
510,000 people out of work, creating the largest industrial closure in Canadian
history and in one of its least-populated provinces. The calamity affected
almost all the half-million inhabitants, including my own family.

Simultaneously, recreational fisheries elsewhere in the world sputtered
down, ending the money that wealthy tourists injected into localities to catch
once abundant "trophy fish" that are rarely or never seen today. Today's trophy
winners might have been laughed at 50 years ago. Increasingly, managers
are finding that species beyond those they target must be considered, to
understand why some years are poor and others are good for fishermen.
This new "ecosystem-based management" recognizes that the predators
(including humans) and prey, food web components, and competitors in
the rich diversity of the ocean all play roles as fisheries collapse, and
sometimes recover.

A diver or snorkeler on coral reefs encounters spectacular beauty and
variety (Figure 2.1). They sense special diversity of life on reefs. Coral reefs
require clear, sunlit shallow areas in tropical latitudes, but more species
have evolved on them than in other marine ecoregions.

Humanity continually interacts with life in temperate and tropical coastal
waters, collecting food, shipping freight, swimming, and gazing at seascapes.

Figure 2.2 Productive kelp forests, filmed without disturbing the inhabitants.
By recycling air exhaled by the divers, the Galatée film team avoided bubbles that might
scare away residents. The divers could also remain underwater for hours by recycling air.
Here, they film in California kelp forests. The Census included these mid-latitude
forests for their abundant and diverse species. See Chapter 5.

Sadly, many activities threaten the beauty and diversity of coral reefs and their function as natural breakwaters and coastal fisheries, putting them at greater risk than any ocean habitat on Planet Ocean. Our surprising lack of knowledge about life in these accessible reefs is an opportunity for discovery.

The cold polar oceans

The understandable ignorance of life in frigid waters opens exciting exploration. For collaborating Census explorers, the Arctic and Antarctic waters at opposite poles have similarities, but also contrasts. Ice generally covers polar regions in winter, shading life beneath it, even as endless nights give way to endless days (Figure 2.3). As the days lengthen, the ice retreats and sunlight penetrates the ocean. Organisms are adapted to year-round water temperatures that may be as cold as $-1.9\ ^\circ C$ without freezing, thanks to the salt. Exposure to cold water, shading ice, daylong darkness, and limited nutrients creates an unproductive polar ocean that bursts to life for short periods leading up to endless days of midnight sun.

Similarities between the poles largely end there. Polar bears live only in the Arctic and penguins only in the southern hemisphere, meeting only in zoos and comic strips. Continents surround the Arctic Ocean basin and ice covers much of it, whereas the Southern Ocean surrounds the ice-covered Antarctic continent. These contrasts create a relatively shallow Arctic Ocean with a wide continental shelf and basin averaging less than 1,300 meters depth, and with significant freshwater inflow. Currents meander around the edge of the basin and form large circulation loops called *gyres*, which interact with weather to affect organisms from year to year. In contrast, the Antarctic continental shelf is relatively deep and narrow and the steep continental slope plunges onto the abyssal plain. The Southern Ocean surrounding the Antarctic continent circulates around the Earth unobstructed by land. In sailor's terms, this open ocean creates infinite *fetch* (the distance the wind blows over the ocean), producing one of the fastest currents and some of the roughest sailing in the world. Little wonder that adventurers like Joshua Slocum spent weeks trying to pass Cape Horn!

Together these similarities and differences allow more than 6,000 species to live in the Arctic, Antarctic, or both. Although more diverse life is thought to live in the Antarctic region, neither the Arctic nor Antarctic has been well explored, and exploration of each continues to yield new species. Indeed, about half of the 1,400 invertebrate species sampled from the abyssal plain adjacent to the Antarctic were new to science. A warmer climate and some of the largest temperature changes are expected in the polar regions, and will melt ice and change food webs and ocean productivity. If warming alters circulation, these effects could be worsened and circulation of pollutants such as mercury would surely change.

Drifting plankton and swimming nekton

Plankton, including microbes, eggs, larvae, and even jellyfish, drift passively at the mercy of currents. In contrast, *nekton* are animals like fish and whales that can make headway against currents and generally swim wherever they choose. Although fauna is typically more concentrated on the seafloor than in the water above, an array of life spends most or all of their lives in the water, drifting and swimming in a three-dimensional world. There is no parallel on land, where even birds rest on the thin veneer of terrestrial Earth that life occupies. Sunlight energy fuels photosynthesis in phytoplankton to create organic material, which in turn fuels most marine food webs. Small animals, the zooplankton, link the phytoplankton to top-level predators targeted by fisheries. Zooplankton also link, often through fishes, to the larger whales, sea lions, and sea turtles that people love, fearing their decline or even endangerment. The smallest and thus least-known plankton are the *microbes*, the mostly single-celled bacteria and protists. Microbes occur everywhere, surprising explorers who find them in rock 1.6 kilometers deep in solid Earth or at 300 °C in superheated hydrothermal vents. Despite their small size that makes them nearly indistinguishable even under a microscope, microbes provide key services and the ocean would die without them. Microbes fuel some of the most productive fisheries, but they also spread disease, like coliform bacteria, or they paralyze shellfish consumers, as some algae do.

Satellite images of vast blue ocean tempt us to assume that plankton drift and nekton swim freely. Temperature and depth, however, create invisible barriers that block traveling sea life, just as the absence of roads on land blocks our traveling cars. These physical differences of temperature and depth create rich fisheries in some waters and unproductive ones elsewhere. Generally coastal waters, where runoff from land, upwelling, and regeneration of mineral nutrients from the seafloor can feed many fish and other life, have more food than the open ocean that lacks these nutrient sources.

Figure 2.3 A sampling of ice habitats.
Polar ice provides habitat for some species despite the cold, and the covering ice
defines the ocean and seafloor habitats for those beneath it. See Chapter 6. **(a)** A scuba
diver's view of Arctic pack ice in the Canada Basin illustrates how the ice can block most
light, but still transmit scattered beams. **(b)** Water fleas, amphipod crustaceans,
feeding on algae that grow under ice near the Beaufort Sea shore. Arctic cod feed
on these ice amphipods, and seals then prey on Arctic cod.

The differences between surface waters near the coast versus the
open ocean affect movement and thus distribution of large nekton such
as whales and seals, sharks, sea turtles, seabirds, and large fishes. Some migrate
seasonally to feed or reproduce, sometimes crossing entire oceans. Some
seabirds circumnavigate the globe. New tracking tools, many pioneered
by the Census, are clarifying why they migrate along specific routes and
whether they migrate to feed or mate.

Migrants like tuna that travel many kilometers for months respect no international boundaries. No nation can regulate these migrants, obviating national management and conservation. Like their counterparts in Newfoundland, tuna fishermen suffer along with their prey, although in this case it is a high-stakes game where each fish may fetch tens of thousands of US dollars. But the new tools for tracking animals offer foundations for logical management and conservation and thus less decline in fish and suffering by fishermen. Moreover, attaching small sensors to animals such as elephant seals that travel far and dive deep in the Antarctic, enables a new oceanography where organisms collect oceanographic data even in the open ocean far from shore.

The open ocean and deep sea

Scientists call the open ocean the *pelagic zone*. It begins at the edge of the continental shelf near 200 meters depth and extends across the open ocean. Because the open ocean covers two-thirds of the Earth, its underlying continental slopes and abyssal plains comprise by far the broadest habitat on Earth, and the open ocean itself provides most of the livable habitat volume on Earth. Within the open ocean, geography, geology, physics, and biology create distinct habitats.

The continental slope is closer to land masses so land runoff, much flowing down rivers to the sea, has greater influence than in the rest of the oceanic realm. During glacial periods, when sea level was lower, rivers carved canyons, which join with other topography, ocean currents, and different productivity of the waters above to create complicated seascapes on continental margins. Fishermen catch grenadiers and Greenland halibut, among others, from the upper continental slope, but deep-sea trawlers must travel far, lose time lowering and recovering nets, and chase scarce fish. Deep-sea fisheries were too costly for most countries, but now the collapse of coastal fisheries and subsidized fishing encourage fishermen to venture into deeper and deeper water.

Sunlight does not penetrate onto most of the continental slope or the abyssal plains, where temperature and salinity vary little. Because darkness means no photosynthesis, the only food falls down from the sunlit surface, often passing through the digestive tracts of mid-water organisms. As it sinks, it is degraded further by bacteria. This food is poetically described as *marine snow*, bits of organic matter sinking from surface waters above. Bottom currents may supply decaying kelp and similar food from productive areas of the continental shelf, but less and less food snows down as distance from land increases.

The pressure of water is immense, reaching several hundred times surface pressure on the abyssal plains. As animals with lungs or sinuses filled with air dive to feed, they must cope with these pressures. If a submersible hull failed, this high pressure would crush passengers before they had a chance to drown! Because pressure modifies how some molecules work, the specialized *enzymes* that facilitate chemical reactions in deep-sea organisms differ from those evolved in shallow water.

High pressure, darkness, and no photosynthesis create an inhospitable environment for life. Thus, the deep ocean was once thought to be a desert, devoid of life. Indeed, early photographs of the deep sea revealed vast plains of gently rolling seafloor covered in sediment (Figure 2.4), propagating the analogy of the abyssal desert that is just now disappearing from textbooks. Today we know that many species live within the sediments, spanning invertebrates, such as polychaete worms, nematodes (thread worms), small clams and snails, sea stars and brittle stars, and shrimp-like crustaceans. They move among the sediment grains and feed on bacteria and bits of degraded food that sink from above. Limited food may limit abundance, but it does not limit diversity. Indeed, far from it.

Exploration of the deep reveals more than rolling plains of sediments. Protruding bedrock provides homes for deep-water corals and other species that require hard surfaces instead of shifting sediment. Large beds of deep-water corals, first discovered off the coast of Norway, form beautiful, dense stands that are important habitat for deep-water species. Census studies around the world on underwater mountains called *seamounts* find corals too, some sparse and some dense. Unlike corals that build reefs in the shallow tropics, corals in the deep, dark ocean lack photosynthetic organisms in their tissue, and instead must feed on passing marine snow and small animals.

As many as 40,000 seamounts rise 1,000 meters from the seafloor, and 200,000 rise more than 100 meters above the abyssal plain. Those that rise above the water can form chains like the Hawaiian Islands. *Mid-ocean ridges* are long, submarine mountain chains that extend thousands of miles through the ocean, rising from joints between the plates covering the Earth, as magma flows and spreads to form new crust. The best-known underwater mountain chain, the Mid-Atlantic Ridge, stretches 10,000 kilometers from Antarctica

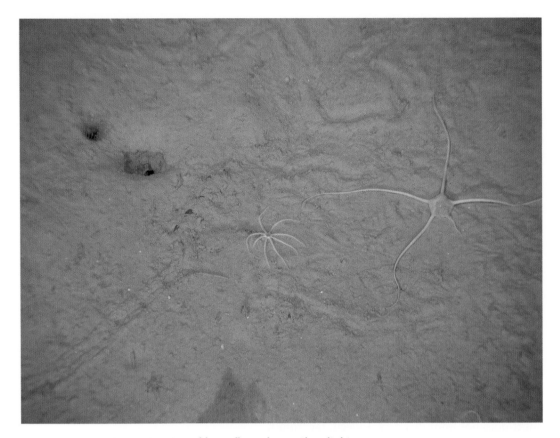

Figure 2.4 A muddy seafloor, deeper than light.
Below 1,000 meters near Nova Scotia, Canada, this type of seascape on a continental slope is the most widespread habitat on Planet Ocean and inspired early analogies to an underwater "desert" long since disproved. See Chapter 8.

to Africa, and north past Greenland into the Arctic. Occasionally it protrudes above the ocean's surface to form islands such as the Azores and Iceland. The Azores extend about 5,000 meters above the abyssal plains that surround them, which does not quite match the height of Mount Everest at 8,848 meters, but does illustrate the scale of ocean ridges.

Fish-finding echosounders led fishermen to seamounts, where nutrient upwelling, accelerating currents, and recirculation attract fish. Seamounts are volcanic in origin and steep sides shed sediment, exposing hard substrate. Cold-water corals and other attached organisms trap food as water flows past and build three-dimensional habitats. Fishermen catch abundant fish around seamounts. Seamounts sometimes support different species than the surrounding abyssal plains, and some chains of seamounts have species not found anywhere else. Census research on the Mid-Atlantic Ridge suggests little *endemism*, where species are unique to that location. Nonetheless, seamount populations may sometimes evolve in isolation from similar populations at distant seamounts, much like the finches on the isolated Galápagos Islands illustrated the idea of evolution in Charles Darwin's *Origin of Species*. Because so few have been sampled, more exploration of seamounts and mid-ocean ridges would likely bring much unknown into the realm of known.

Seamounts create isolated mountains in the otherwise muddy, rolling hills of the abyssal plains and thus support specialized deep-sea faunas. Their depth and distance from land makes scientific expeditions difficult to mount. But recognizing them as a frontier rarely explored, and a laboratory for understanding ecosystems scarcely altered by humans, scientists are leading a new surge of interest.

Small hydrothermal vents first discovered in 1977 flow at the edges of tectonic plates. While the edges of some plates ooze and spread new crust, others collide, and one plate is pushed beneath the other in a process called *subduction*. Subduction destroys the old ocean crust that moves deep into the Earth. During spreading and subduction the ocean crust cracks, letting seawater percolate into the upper mantle, where it is superheated and enriched with metals and other compounds. When the fluid reemerges into the ocean, it can be 300–400 °C and rich in hydrogen sulfide, methane, carbon monoxide, and metal ions that are toxic to much marine life. The discovery of a huge biomass (kilograms per square meter) of meter-long tube worms and giant clams in the seemingly toxic waters around vents (Figure 2.5) astounded scientists because, with a few exceptions, the deep sea typically has few small organisms. An American gossip magazine known for its exaggeration once interviewed Fred Grassle, but the story of hydrothermal vents was so fantastic that they reported it accurately without embellishment. Truth stranger than fiction! But the subsequent discovery that the giant worms and clams were made possible by bacteria living inside them that utilize chemical energy in the vent fluid excited scientists even more. This discovery, made decades ago, ended the view that all ecosystems are fueled by photosynthesis. In vent ecosystems, specialized bacteria and archaea use chemical instead of sunlight energy to create organic molecules that are the basis of life. Ocean *cold seeps*, where methane, hydrocarbons, and hydrogen sulfide slowly bubble from the seafloor and fuel bacterial production, also support a large biomass of organisms.

Figure 2.5 An oasis of the deep.
Zoarcid fish, or eelpouts, swim over a community of tubeworms, *Riftia pachyptila*, on an East Pacific hydrothermal vent. The huge biomass of tubeworms and other species near hydrothermal vents contrasts starkly with the sparse biomass in muddy sediments under much of the deep sea. These abundances earned hydrothermal vents their nickname of "underwater oases."

During the 30 years of study since the discovery of vents and seeps, more than 700 species have been discovered at hydrothermal vents, and 600 species at seeps. Carcasses of whales that fall to the seafloor provide potential stepping stones between vents and seeps, but also support species unknown elsewhere. Even during the first decade of the twenty-first century, long after the Age of Discovery, Census explorers found new vents in the Arctic and Southern Ocean, moving unknown into known and raising the number of known species, a prelude to the riot of species in the many environments of Planet Ocean.

BIBLIOGRAPHY

Amaral-Zettler, L., L. F. Artigas, J. Baross, *et al.*, A global census of marine microbes. In A. D. McIntyre (ed.), *Life in the World's Oceans: Diversity, Distribution, and Abundance* (Oxford: Blackwell Publishing Ltd., 2010), pp. 223–45.

Bailey, H., G. Shillinger, D. Palacios, *et al.*, Identifying and comparing phases of movement by leatherback turtles using state-space models. *J. Exp. Mar. Biol. Ecol.*, **356**:1–2 (2008), 128–35.

Baker, M. C., E. Z. Ramirez-Llodra, P. A. Tyler, *et al.*, Biogeography, ecology and vulnerability of chemosynthetic ecosystems in the deep sea. In A. D. McIntyre (ed.), *Life in the World's Oceans: Diversity, Distribution, and Abundance* (Oxford: Blackwell Publishing Ltd., 2010), pp. 161–82.

Balata, D., I. Bertocci, L. Piazzi, and U. Nesti, Comparison between epiphyte assemblages of leaves and rhizomes of the seagrass *Posidonia oceanica* subjected to different levels of anthropogenic eutrophication. *Estuarine Coastal Shelf Sci.*, **79**:3 (2008), 533–40.

Benedetti-Cecchi, L., Understanding the consequences of changing biodiversity on rocky shores: how much have we learned from past experiments? *J. Exp. Mar. Biol. Ecol.*, **338**:2 (2006), 193–204.

Benedetti-Cecchi, L., I. Bertocci, S. Vaselli, and E. Maggi, Temporal variance reverses the impact of high mean intensity of stress in climate change experiments. *Ecology*, **87**:10 (2006), 2489–99.

Benedetti-Cecchi, L., E. Maggi, I. Bertocci, *et al.*, Variation in rocky shore assemblages in the northwestern Mediterranean: contrasts between islands and the mainland. *J. Exp. Mar. Biol. Ecol.*, **293**:2 (2003), 193–215.

Benedetti-Cecchi, L. and G. C. Osio, Replication and mitigation of effects of confounding variables in environmental impact assessment: effect of marinas on rocky-shore assemblages. *Mar. Ecol. Prog. Ser.*, **334** (2007), 21–35.

Bergstad, O. A. and O. R. Godø, The pilot project "Patterns and processes of the ecosystems of the northern Mid-Atlantic": aims, strategy and status. *Oceanol. Acta*, **25**:5 (2002), 219–26.

Bertocci, I., E. Maggi, S. Vaselli, and L. Benedetti-Cecchi, Contrasting effects of mean intensity and temporal variation of disturbance on a rocky seashore. *Ecology*, **86**:8 (2005), 2061–7.

Bertocci, I., S. Vaselli, E. Maggi, and L. Benedetti-Cecchi, Changes in temporal variance of rocky shore organism abundances in response to manipulation of mean intensity and temporal variability of aerial exposure. *Mar. Ecol. Prog. Ser.*, **338** (2007), 11–20.

Block, B. A., Physiological ecology in the 21st century: advancements in biologging science. *Integr. Comp. Biol.*, **45**:2 (2005), 305–20.

Block, B. A., H. Dewar, S. B. Blackwell, *et al.*, Migratory movements, depth preferences, and thermal biology of Atlantic bluefin tuna. *Science*, **293**:5533 (2001), 1310–4.

Block, B. A., H. Dewar, C. J. Farwell, and E. D. Prince, A new satellite technology for tracking the movements of Atlantic bluefin tuna. *Proc. Natl. Acad. Sci. USA*, **95** (1998), 9384–9.

Block, B. A., S. L. H. Teo, A. Walli, *et al.*, Electronic tagging and population structure of Atlantic bluefin tuna. *Nature*, **434**:7037 (2005), 1121–7.

Bluhm, B. A., I. R. MacDonald, C. Debenham, and K. Iken, Macro- and megabenthic communities in the high Arctic Canada Basin: initial findings. *Polar Biol.*, **28**:3 (2005), 218–31.

Boehlert, G. W., D. P. Costa, D. E. Crocker, *et al.*, Autonomous pinniped environmental samplers: using instrumented animals as oceanographic data collectors. *J. Atmos. Oceanic Technol.*, **18**:11 (2001), 1882–93.

Bortolus, A., O. O. Iribarne, and M. M. Martinez, Relationship between waterfowl and the seagrass *Ruppia maritima* in a southwestern Atlantic coastal lagoon. *Estuaries*, **21** (1998), 710–7.

Botsford, L. W., J. C. Castilla, and C. H. Peterson, The management of fisheries and marine ecosystems. *Science*, **277**:5325 (1997), 509–15.

Boustany, A. M., S. F. Davis, P. Pyle, *et al.*, Satellite tagging – expanded niche for white sharks. *Nature*, **415**:6867 (2002), 35–6.

Brandt, A., C. De Broyer, I. De Mesel, *et al.*, The biodiversity of the deep Southern Ocean benthos. *Philos. Trans. R. Soc. Lond. B Biol. Sci.*, **362**:1477 (2007), 39–66.

Brandt, A., A. J. Gooday, S. N. Brandao, *et al.*, First insights into the biodiversity and biogeography of the Southern Ocean deep sea. *Nature*, **447**:7142 (2007), 307–11.

Bucklin, A., S. Nishida, S. Schnack-Schiel, *et al.*, A census of the zooplankton of the global ocean. In A. D. McIntyre (ed.), *Life in the World's Oceans: Diversity, Distribution, and Abundance* (Oxford: Blackwell Publishing Ltd., 2010), pp. 247–65.

Bulleri, F. and L. Benedetti-Cecchi, Facilitation of the introduced green alga *Caulerpa racemosa* by resident algal turfs: experimental evaluation of underlying mechanisms. *Mar. Ecol. Prog. Ser.*, **364** (2008), 77–86.

Butman, C., J. T. Carlton, and S. R. Palumbi, Whaling effects on deep-sea biodiversity. *Conserv. Biol.*, **9**:2 (1995), 462–4.

Carlton, J. T. and J. B. Geller, Ecological roulette – the global transport of nonindigenous marine organisms. *Science*, **261**:5117 (1993), 78–82.

Cavanaugh, C. M., S. L. Gardiner, M. L. Jones, H. W. Jannasch, and J. B. Waterbury, Prokaryotic cells in the hydrothermal vent tube worm *Riftia pachyptila* Jones: possible chemoautotrophic symbionts. *Science*, **213**:4505 (1981), 340–2.

Chapman, P. M. and M. J. Riddle, Toxic effects of contaminants in polar marine environments. *Environ. Sci. Technol.*, **39**:9 (2005), 200A–7A.

Chenelot, H., K. Iken, B. Konar, and M. Edwards, Spatial and temporal distribution of echinoderms in rocky nearshore areas of Alaska. In P. R. Rigby and Y. Shirayama (eds.), *Selected Papers of the NaGISA World Congress 2006*, Publications of the Seto Marine Biological Laboratory, Special Publication Series Vol. VIII (Kyoto, Japan: Kyoto University, 2007), pp. 11–28.

Chenelot, H. and B. Konar, *Lacuna vincta* (Mollusca, Neotaenioglossa) herbivory on juvenile and adult *Nereocystis luetkeana* (Heterokontophyta, Laminariales). *Hydrobiologia*, **583** (2007), 107–18.

Clark, M. R. and A. A. Rowden, Effect of deepwater trawling on the macro-invertebrate assemblages of seamounts on the Chatham Rise, New Zealand. *Deep Sea Res. I*, **56**:9 (2009), 1540–54.

Clark, M. R., D. P. Tittensor, A. D. Rogers, *et al.*, *Seamounts, Deep-sea Corals and Fisheries: Vulnerability of Deep-sea Corals to Fishing on Seamounts Beyond Areas of National Jurisdiction* (Cambridge, UK: UNEP-WCMC, 2006).

Clark, M. R., V. I. Vinnichenko, J. D. M. Gordon, *et al.*, Large-scale distant-water trawl fisheries on seamounts. In T. J. Pitcher, T. Morato, P. J. B. Hart, M. R. Clark, N. Haggan, and R. S. Santos (eds.), *Seamounts: Ecology, Fisheries & Conservation* (Oxford: Wiley-Blackwell, 2007), pp. 361–99.

Consalvey, M., M. R. Clark, A. A. Rowden, and K. I. Stocks, Life on seamounts. In A. D. McIntyre (ed.), *Life in the World's Oceans: Diversity, Distribution, and Abundance* (Oxford: Blackwell Publishing Ltd., 2010), pp. 123–38.

Cooke, S. J., S. G. Hinch, A. P. Farrell, *et al.*, Developing a mechanistic understanding of fish migrations by linking telemetry with physiology, behavior, genomics and experimental biology: an interdisciplinary case study on adult Fraser River sockeye salmon. *Fisheries*, **33**:7 (2008), 321–38.

Costa, D. P. and B. Sinervo, Field physiology: physiological insights from animals in nature. *Annu. Rev. Physiol.*, **66** (2004), 209–38.

Croxall, J. P., J. R. D. Silk, R. A. Phillips, V. Afanasyev, and D. R. Briggs, Global circumnavigations: tracking year-round ranges of nonbreeding albatrosses. *Science*, **307**:5707 (2005), 249–50.

Cunningham, L., I. Snape, J. S. Stark, and M. J. Riddle, Benthic diatom community response to environmental variables and metal concentrations in a contaminated bay adjacent to Casey Station, Antarctica. *Mar. Pollut. Bull.*, **50**:3 (2005), 264–75.

Cunningham, L., J. S. Stark, I. Snape, A. McMinn, and M. J. Riddle, Effects of metal and petroleum hydrocarbon contamination on benthic diatom communities near Casey Station, Antarctica: an experimental approach. *J. Phycol.*, **39**:3 (2003), 490–503.

Desbruyeres, D., N. Segonzac, and M. Bright, *Handbook of Deep-sea Hydrothermal Vent Fauna*, 2nd edn (Denisia: Linz, 2006).

Dewar, H., M. Domeier, and N. Nasby-Lucas, Insights into young of the year white shark, *Carcharodon carcharias*, behavior in the Southern California Bight. *Environ. Biol. Fishes*, **70**:2 (2004), 133–43.

Ebbe, B., D. Billett, A. Brandt, *et al.*, Diversity of abyssal marine life. In A. D. McIntyre (ed.), *Life in the World's Oceans: Diversity, Distribution, and Abundance* (Oxford: Blackwell Publishing Ltd., 2010), pp. 139–60.

Estes, J. A. and D. O. Duggins, Sea otters and kelp forests in Alaska: generality and variation in a community ecological paradigm. *Ecol. Monogr.*, **65**:1 (1995), 75–100.

Forbes, E., *Report on the Mollusca and Radiata of the Aegean Sea*. 1844.

de Forges, B. R., J. A. Koslow, and G. C. B. Poore, Diversity and endemism of the benthic seamount fauna in the southwest Pacific. *Nature*, **405**:6789 (2000), 944–7.

Fossa, J. H., P. B. Mortensen, and D. M. Furevik, The deep-water coral *Lophelia pertusa* in Norwegian waters: distribution and fishery impacts. *Hydrobiologia*, **471** (2002), 1–12.

Gagaev, S. Y., *Sigambra healyae sp n.*, a new species of polychaete (Polychaeta: Pilargidae) from the Canadian Basin of the Arctic Ocean. *Russ. J. Mar. Biol.*, **34**:1 (2008), 73–5.

Gien, L. T., Land and sea connection: the East Coast fishery closure, unemployment and health. *Can. J. Public. Health.*, **91**:2 (2000), 121–4.

Gradinger, R. and B. Bluhm, Arctic ocean exploration 2002. *Polar Biol.*, **28**:3 (2005), 169–70.

Gradinger, R., B. A. Bluhm, R. R. Hopcroft, *et al.*, Marine life in the Arctic. In A. D. McIntyre (ed.), *Life in the World's Oceans: Diversity, Distribution, and Abundance* (Oxford: Blackwell Publishing Ltd., 2010), pp. 183–202.

Grassle, J. F., The ecology of deep-sea hydrothermal vent communities. *Adv. Mar. Biol.* (1986), 301–62.

Greene, C. H., B. A. Block, D. Welch, *et al.*, Advances in conservation oceanography: new tagging and tracking technologies and their potential for transforming the science underlying fisheries management. *Oceanography*, **22**:1 (2009), 210–23.

Greene, C. H., A. J. Pershing, T. M. Cronin, and N. Ceci, Arctic climate change and its impacts on the ecology of the North Atlantic. *Ecology*, **89**:11 (2008), S24–S38.

Gutt, J., G. Hosie, and M. Stoddart, Marine life in the Antarctic. In A. D. McIntyre (ed.), *Life in the World's Oceans: Diversity, Distribution, and Abundance* (Oxford: Blackwell Publishing Ltd., 2010), pp. 203–20.

Gutt, J., B. Sirenko, I. Smirnov, and W. Arntz, How many macrobenthic species might inhabit the Antarctic shelf? *Antarct. Sci.*, **16** (2004), 11–6.

Haury, L., C. Fey, C. Newland, and A. Genin, Zooplankton distribution around four eastern North Pacific seamounts. *Prog. Oceanogr.*, **45**:1 (2000), 69–105.

Hillier, J. K. and A. B. Watts, Global distribution of seamounts from ship-track bathymetry data. *Geophys. Res. Lett.*, **34**:13 (2007).

Hinrichsen, D. *Coasts in Crisis* (1995). Available from: http://www.aaas.org/international/ehn/fisheries/hinrichs.htm.

Hutchings, J. A., Spatial and temporal variation in the density of northern cod and a review of hypotheses for the stock's collapse. *Can. J. Fish. Aquat. Sci.*, **53**:5 (1996), 943–62.

Incze, L. S., P. Lawton, S. L. Ellis, and N. H. Wolff, Biodiversity knowledge and its application in the Gulf of Maine area. In A. D. McIntyre (ed.), *Life in the World's Oceans: Diversity, Distribution, and Abundance* (Oxford: Blackwell Publishing Ltd., 2010), pp. 43–63.

Kitagawa, T., A. M. Boustany, C. J. Farwell, *et al.*, Horizontal and vertical movements of juvenile bluefin tuna (*Thunnus orientalis*) in relation to seasons and oceanographic conditions in the eastern Pacific Ocean. *Fish Oceanogr.*, **16**:5 (2007), 409–21.

Knowlton, N., Coral reef coda: what can we hope for? In I. Cote and J. Reynolds (eds.), *Coral Reef Conservation* (Cambridge, UK: Cambridge University Press, 2006), pp. 538–49.

Knowlton, N., R. E. Brainard, R. Fisher, *et al.*, Coral reef biodiversity. In A. D. McIntyre (ed.), *Life in the World's Oceans: Diversity, Distribution, and Abundance* (Oxford: Blackwell Publishing Ltd., 2010), pp. 65–77.

Knowlton, N. and J. B. C. Jackson, Shifting baselines, local impacts, and global change on coral reefs. *PLoS Biol.*, **6**:2 (2008), e54.

Konar, B., Recolonization of a high latitude hard-bottom nearshore community. *Polar Biol.*, **30**:5 (2007), 663–7.

Konar, B., K. Iken, and M. Edwards, Depth-stratified community zonation patterns on Gulf of Alaska rocky shores. *Mar. Ecol.*, **30**:1 (2009), 63–73.

Konar, B., R. Riosmena-Rodriguez, and K. Iken, Rhodolith bed: a newly discovered habitat in the North Pacific Ocean. *Bot. Mar.*, **49**:4 (2006), 355–9.

Koslow, J. A., G. W. Boehlert, J. D. M. Gordon, *et al.*, Continental slope and deep-sea fisheries: implications for a fragile ecosystem. *ICES J. Mar. Sci.*, **57**:3 (2000), 548–57.

Lajus, D. L., Z. V. Dmitrieva, A. V. Kraikovski, J. A. Lajus, and D. A. Alexandrov, Atlantic salmon fisheries in the White and Barents Sea basins: dynamic of catches in the 17–18th century and comparison with 19–20th century data. *Fish. Res.*, **87**:2–3 (2007), 240–54.

Laurel, B. J., R. S. Gregory, and J. A. Brown, Settlement and distribution of Age-0 juvenile cod, *Gadus morhua* and *G. ogac*, following a large-scale habitat manipulation. *Mar. Ecol. Prog. Ser.*, **262** (2003), 241–52.

Le Boeuf, B. J., D. E. Crocker, J. Grayson, *et al.*, Respiration and heart rate at the surface between dives in northern elephant seals. *J. Exp. Biol.*, **203**:21 (2000), 3265–74.

Levin, L. A., D. F. Boesch, A. Covich, *et al.*, The function of marine critical transition zones and the importance of sediment biodiversity. *Ecosystems*, **4**:5 (2001), 430–51.

Lotze, H. K., Rise and fall of fishing and marine resource use in the Wadden Sea, southern North Sea. *Fish. Res.*, **87**:2–3 (2007), 208–18.

Lotze, H. K. and M. Glaser, Ecosystem services of semi-enclosed marine systems. In E. R. Urban, B. Sundby, P. Malanotte-Rizzoli, and J. M. Melillo (eds.), *Watersheds, Bays and Bounded Seas* (Washington, DC: Island Press, 2008), pp. 227–49.

Lotze, H. K., K. Reise, B. Worm, *et al.*, Human transformations of the Wadden Sea ecosystem through time: a synthesis. *Helgol. Mar. Res.*, **59**:1 (2005), 84–95.

Lotze, H. K. and B. Worm, Historical baselines for large marine animals. *Trends Ecol. Evol.*, **24**:5 (2009), 254–62.

MacKenzie, B. R. and R. A. Myers, The development of the northern European fishery for north Atlantic bluefin tuna *Thunnus thynnus* during 1900–1950. *Fish. Res.*, **87**:2–3 (2007), 229–39.

McClain, C. R., Seamounts: identity crisis or split personality? *J. Biogeogr.*, **34**:12 (2007), 2001–8.

McClenachan, L., Documenting loss of large trophy fish from the Florida Keys with historical photographs. *Conserv. Biol.*, **23**:3 (2009), 636–43.

McKenzie, M. G. and G. Matthew, Baiting our memories: the impact of offshore technology change on the species around Cape Cod, 1860–1895. In D. Starckey, P. Holm and M. Barnard (eds.), *Oceans Past: Management Insights from the History of Marine Animal Populations* (London: EarthScan Press, 2007), pp. 77–89.

Menot, L., M. Sibuet, R. S. Carney, *et al.*, New perceptions of continental margin biodiversity. In A. D. McIntyre (ed.), *Life in the World's Oceans: Diversity, Distribution, and Abundance* (Oxford: Blackwell Publishing Ltd., 2010), pp. 79–101.

Miloslavich, P. and E. Klein (eds.), *Caribbean Marine Biodiversity: The Known and Unknown* (Lancaster, PA: DEStech Publications, 2005).

Morato, T., W. W. L. Cheung, and T. Pitcher, Additions to Froese and Sampang's checklist of seamount fishes. In T. Morato and D. Pauly (eds.), *Seamounts: Biodiversity and Fisheries* (Oxford: Blackwell Publishing, 2004), pp. 51–60.

Myers, R. A. and B. Worm, Rapid worldwide depletion of predatory fish communities. *Nature*, **423**:6937 (2003), 280–3.

Myers, R. A. and B. Worm, Extinction, survival or recovery of large predatory fishes. *Philos. Trans. R. Soc. Lond. B Biol. Sci.*, **360**:1453 (2005), 13–20.

Naylor, R. L., R. J. Goldburg, H. Mooney, *et al.*, Nature's subsidies to shrimp and salmon farming. *Science*, **282**:5390 (1998), 883–4.

O'Dor, R. K., K. Fennel, and E. Vanden Berghe, A one ocean model of biodiversity. *Deep Sea Res. II*, **56**:19–20 (2009), 1816–23.

O'Hara, T. D., A. A. Rowden, and A. Williams, Cold-water coral habitats on seamounts: do they have a specialist fauna? *Divers. Distrib.*, **14**:6 (2008), 925–34.

Ojaveer, H., K. Awebro, H. M. Karlsdottir, and B. R. MacKenzie, Swedish Baltic Sea fisheries during 1868–1913: spatio-temporal dynamics of catch and fishing effort. *Fish. Res.*, **87**:2–3 (2007), 137–45.

Palacios, D. M. and S. J. Bograd, A census of Tehuantepec and Papagayo eddies in the northeastern tropical Pacific. *Geophys. Res. Lett.*, **32**:23 (2005).

Palacios, D. M., S. J. Bograd, D. G. Foley, and F. B. Schwing, Oceanographic characteristics of biological hot spots in the North Pacific: A remote sensing perspective. *Deep Sea Res. II*, **53**:3–4 (2006), 250–69.

Pandolfi, J. M., R. H. Bradbury, E. Sala, *et al.*, Global trajectories of the long-term decline of coral reef ecosystems. *Science*, **301**:5635 (2003), 955–8.

Pandolfi, J. M., J. B. C. Jackson, N. Baron, *et al.*, Ecology – are US coral reefs on the slippery slope to slime? *Science*, **307**:5716 (2005), 1725–6.

Piraino, S., B. A. Bluhm, R. Gradinger, and F. Boero, *Sympagohydra tuuli* gen. nov and sp nov (Cnidaria: Hydrozoa) a cool hydroid from the Arctic sea ice. *J. Mar. Biol. Assoc. UK*, **88**:8 (2008), 1637–41.

Ramirez-Llodra, E., M. Blanco, and A. Arcas, ChEssBase: an online information system on biodiversity and biogeography of deep-sea chemosynthetic ecosystems. 2004; Available from: www.noc.soton.ac.uk/chess/db_home.php.

Rasmussen, K., D. M. Palacios, J. Calambokidis, *et al.*, Southern Hemisphere humpback whales wintering off Central America: insights from water temperature into the longest mammalian migration. *Biol. Lett.*, **3**:3 (2007), 302–5.

Reed, C., Marine science: boiling points. *Nature*, **439**:7079 (2006), 905–7.

Rosenberg, A. A., W. J. Bolster, K. E. Alexander, *et al.*, The history of ocean resources: modeling cod biomass using historical records. *Front. Ecol. Environ.*, **3**:2 (2005), 84–90.

Ross, D. A., *Introduction to Oceanography* (New York: Harper Collins College Publishers, 1995).

Shaffer, S. A., Y. Tremblay, H. Weimerskirch, *et al.*, Migratory shearwaters integrate oceanic resources across the Pacific Ocean in an endless summer. *Proc. Natl. Acad. Sci. USA*, **103**:34 (2006), 12799–802.

Shillinger, G. L., D. M. Palacios, H. Bailey, *et al.*, Persistent leatherback turtle migrations present opportunities for conservation. *PLoS Biol.*, **6**:7 (2008), 1408–16.

Smith, C. R. and A. R. Baco, Ecology of whale falls at the deep-sea floor. *Oceanogr. Mar. Biol. Ann. Rev.*, **41** (2003), 311–54.

Smith, K. L., H. A. Ruhl, R. S. Kaufmann, and M. Kahru, Tracing abyssal food supply back to upper-ocean processes over a 17-year time series in the northeast Pacific. *Limnol. Oceanogr.*, **53**:6 (2008), 2655–67.

Snelgrove, P. V. R., T. H. Blackburn, P. A. Hutchings, *et al.*, The importance of marine sediment biodiversity in ecosystem processes. *Ambio*, **26**:8 (1997), 578–83.

Snelgrove, P. V. R., M. Flitner, E. R. Urban, *et al.*, Governance and management of ecosystem services in semi-enclosed marine systems. In E. R. Urban, B. Sundby, P. Malanotte-Rizzoli, and J. M. Melillo (eds.), *Watersheds, Bays and Bounded Seas* (Washington, DC: Island Press, 2008), pp. 49–76.

Snelgrove, P. V. R. and C. R. Smith, A riot of species in an environmental calm: the paradox of the species-rich deep-sea floor. *Oceanogr. Mar. Biol. Ann. Rev.*, **40** (2002), 311–42.

Somero, G. N., Environmental adaptation of proteins: strategies for the conservation of critical functional and structural traits. *Comp. Biochem. Physiol. A Comp. Physiol.*, **76**:3 (1983), 621–33.

Spalding, M. D., H. E. Fox, G. R. Allen, *et al.*, Marine ecoregions of the world: a bioregionalization of coastal and shelf areas. *Bioscience*, **57**:7 (2007), 573–83.

Stemmann, L., A. Hosia, M. J. Youngbluth, *et al.*, Vertical distribution (0–1000 m) of macrozooplankton, estimated using the Underwater Video Profiler, in different hydrographic regimes along the northern portion of the Mid-Atlantic Ridge. *Deep Sea Res. II*, **55**:1–2 (2008), 94–105.

Stempniewicz, L., K. Blachowlak-Samolyk, and J. M. Weslawski, Impact of climate change on zooplankton communities, seabird populations and arctic terrestrial ecosystem – a scenario. *Deep Sea Res. II*, **54**:23–26 (2007), 2934–45.

Teo, S. L. H., A. Boustany, H. Dewar, *et al.*, Annual migrations, diving behavior, and thermal biology of Atlantic bluefin tuna, *Thunnus thynnus*, on their Gulf of Mexico breeding grounds. *Mar. Biol.*, **151**:1 (2007), 1–18.

Turner, R. E. and N. N. Rabalais, Linking landscape and water quality in the Mississippi River Basin for 200 years. *Bioscience*, **53**:6 (2003), 563–72.

Van Dover, C. L., S. E. Humphris, D. Fornari, *et al.*, Biogeography and ecological setting of Indian Ocean hydrothermal vents. *Science*, **294**:5543 (2001), 818–23.

Vaselli, S., I. Bertocci, E. Maggi, and L. Benedetti-Cecchi, Effects of mean intensity and temporal variance of sediment scouring events on assemblages of rocky shores. *Mar. Ecol. Prog. Ser.*, **364** (2008), 57–66.

Vecchione, M., O. A. Bergstad, I. Byrkjedal, *et al.*, Biodiversity patterns and processes on the Mid-Atlantic Ridge. In A. D. McIntyre (ed.), *Life in the World's Oceans: Diversity, Distribution, and Abundance* (Oxford: Blackwell Publishing Ltd., 2010), pp. 103–21.

Wagey, T. and Z. Arifin, *Marine Biodiversity Review of the Arafura and Timor Seas* (Ministry of Marine Affairs and Fisheries, Indonesian Institute of Sciences, United Nations Development Program, and Census of Marine Life, 2008).

Wassmann, P., M. Reigstad, T. Haug, *et al.*, Food webs and carbon flux in the Barents Sea. *Prog. Oceanogr.*, **71**:2–4 (2006), 232–87.

Waycotta, M., C. M. Duarte, T. J. B. Carruthers, *et al.*, Accelerating loss of seagrasses across the globe threatens coastal ecosystems. *Proc. Natl. Acad. Sci. USA*, **106**:30 (2009), 12377–81.

Welch, D. W., G. W. Boehlert, and B. R. Ward, POST – the Pacific Ocean Salmon Tracking project. *Oceanol. Acta*, **25**:5 (2002), 243–53.

Welch, D. W., E. L. Rechisky, M. C. Melnychuk, *et al.*, Survival of migrating salmon smolts in large rivers with and without dams. *PLoS Biol.*, **6**:10 (2008), e265.

Whitman, W. B., D. C. Coleman, and W. J. Wiebe, Prokaryotes: the unseen majority. *Proc. Natl. Acad. Sci. USA*, **95**:12 (1998), 6578–83.

Wiencke, C., M. N. Clayton, I. Gomez, *et al.*, Life strategy, ecophysiology and ecology of seaweeds in polar waters. *Rev. Environ. Sci. Biotechnol.*, **6**:1–3 (2007), 95–126.

Worm, B. and H. K. Lotze, Changes in marine biodiversity as an indicator of climate change. In T. Letcher (ed.), *Climate Change: Observed Impacts on Planet Earth* (Amsterdam: Elsevier, 2009), pp. 263–79.

Worm, B., H. K. Lotze, I. Jonsen, and C. Muir, The future of marine animal populations. In A. D. McIntyre (ed.), *Life in the World's Oceans: Diversity, Distribution, and Abundance* (Oxford: Blackwell Publishing Ltd., 2010), pp. 315–30.

Youngbluth, M., T. Sornes, A. Hosla, and L. Stemmann, Vertical distribution and relative abundance of gelatinous zooplankton, in situ observations near the Mid-Atlantic Ridge. *Deep Sea Res. II*, **55**:1–2 (2008), 119–25.

3

A riot of species from microbes to whales

Centuries after the Age of Discovery, scientists still typically describe about four new marine species every day, not even including new microbes. Hydrothermal vents alone yield a newly described species about every two weeks and recently 90% of invertebrate species found in one area of the abyssal plains were new species. Explorers continue to pull new and once "extinct" species from the sea.

Recent discoveries include new fishes, lobsters, bizarre crabs, small crustaceans, octopuses, specialized worms, and carnivorous deep-sea sponges (Figure 1.5). They include sea cucumbers that walk across the seafloor, and a whole array of microbes from the Antarctic to deep-sea hydrothermal vents. Rarely explored habitats from the poles to open ocean waters, to deep-sea sediments, trenches, deep-ocean ridges, and hydrothermal vents added to the total, but only scratched the surface of the unknown species in these places. Skilled explorers are few, so discovery is slowed by how much these few experts can do and not by the large pool of undiscovered life.

Familiar coastal waters or intertidal pools along developed nations scoured by scientists for centuries unexpectedly raise the total of new species even more. For example, in 2006 a customs officer in Durban, South Africa relayed to Census scientists Johan Groeneveld and Charles Griffiths a fisherman's request to export what were labeled as European lobsters. But European lobsters did not occur anywhere near where they had been fishing. The 4-kilogram lobster the fisherman wanted to export were discovered to be new to science, even though South Africa marine fauna is well studied.

Fishes, which are among the best-known marine life, add about 150 species per year. And discovery is not limited to rare species. The most numerous photosynthetic organism in the world, a small single-celled bacterium named *Prochlorococcus*, was unknown until 1985 when electronic "noise" turned out to be an unknown, but widely distributed and abundant species invisible under most microscopes, but detectable with lasers.

We now know that the oceanic riot of species will eventually number in the millions, an immense diversity. During the past 100 years, improved technologies have raised the projected number by making us more efficient explorers. But there is uncertainty. Scientists even debate whether hundreds of thousands or many millions of species live in the ocean.

Unquestionably a greater diversity of broad groups of animals live in the ocean than on land or freshwater. Some animal *phyla*, a broad grouping of similar types of organisms such as the Echinodermata, including sea stars and sea urchins, live throughout the ocean, but not on land or in freshwater. But at the species level, the question of where most species live remains open. We are a terrestrial species that favors exploration of land over sea. The many tropical terrestrial insects, also mostly unknown, might tip the scales toward land. In the sea, mammals are the best-known group, and the fish are also comparatively well known. Nevertheless, new species of fish continue to be found in the Caribbean with no sign of slowing down since the 1700s, when descriptions started to accumulate. Many scientists believe that most marine diversity has not yet been sampled and described.

Just what is a species? Although most people would answer correctly that humans, chimpanzees, and dogs are separate species, the answer becomes harder when all organisms are considered. We typically define species as populations that at least have the potential to interbreed. *Asexual reproduction* in single-celled species that simply divide or others that split off tissue fall outside this definition. Sexual breeding can be difficult to confirm, and is particularly problematic to test among look-alike species. Also, in *sexually dimorphic* species, where males and females appear very different, the sexes are easily interpreted as separate species. Developmental stages such as caterpillars and butterflies on land might be mistaken for different species, and at sea, stages of a single species of jellyfish have frequently been called distinct species. Finally, because of the problem with the typical definition of species, microbiologists now often finesse the problem by referring to *operational taxonomic units*, or OTUs, rather than species. OTUs are genetically distinct and assumed to be different from other OTUs. So what marine diversity is known?

The history of exploration of marine biodiversity

Human naming and drawing of life dates to the earliest recorded communication. Consider the ~4,000-year-old drawings of dolphins on the palace walls at Knossos in Crete, and sturgeon on fifth-century BC Phoenician coins. In about 350 BC, Aristotle recorded and named hundreds of marine animals in his *History of Animals*, so he became the "father of marine biology." Practically, naming animals communicated which could be eaten and which were dangerous. Clues suggest harvesting of shellfish and finfish began in the middle Stone Age. Second-century Latin and Greek verse suggests trawling by Romans, who must have sorted the trawl contents into edible, ornamental, and poisonous.

Uncoordinated naming of which species was which created a Tower of Babel among regions and languages. Different cultures, and sometimes even each tribe and village, called species by a different name. Thus, how could a naturalist collecting codfish off the coast of Cape Cod be certain it was the same species being fished and marketed as torsk in Denmark, dorsz in Poland, and bacalhau in Portugal?

Almost 300 years ago the Swedish botanist, zoologist (and physician) Carl Linnaeus invented *binomial nomenclature*. Wisely avoiding argument about which modern language to use, the Linnaean system speaks in the near-universal Latin language. The system organized the Latin names in a hierarchy according to similar appearance, which Linnaeus attributed to God's greater plan. Fortunately, that similarity usually reflects evolutionary relationships among species. For example, *Gadus morhua*, the binomial name of Atlantic cod, begins with the genus *Gadus* (which is also the genus for several closely related species of cod). The addition of *morhua* forms the unique species name *Gadus morhua*.

During his lifetime, Linnaeus classified many species as he expanded his initial 11 page *Systema Naturae* to more than 3,000 pages in the 13th edition. Understandably, he and many others changed some classifications as new information arrived, particularly about evolutionary relationships. Linnaeus had only the physical appearance or morphology for his classification, and we know now that some species vary in appearance and some genetically separate species look alike. In species where males and females are completely different, physical appearance is almost useless for inferring they are the same species. During the last few decades, some problems were resolved with unique enzymes called *allozymes* and then revolutionary molecular tools, such as barcoding, to rapidly identify species. These new tools built on Linnaeus' work to change biology forever.

Following early waders near the shore, the history of ocean exploration reached Phoenicians, Polynesians, and Vikings, and then European explorers to the New World. Because early exploration sought new lands and sea routes, navigation, coastlines, and currents were of first interest. Nonetheless, the logs of early explorers such as Columbus and Caboto record an abundance of fish and presence of large animals such as whales and seals that suggest a world unlike our modern one. The widespread name Tortuga (meaning turtle) for hundreds of islands, keys, and headlands confirms records in ships' logs of abundant turtles. Delving into such historical information, the Census program *History of Marine Animal Populations* (HMAP) created the new scientific discipline known as environmental history, which recreates the ocean past.

Coordinated scientific investigations on marine expeditions began during 1700–1900. Such expeditions as James Cook's focused primarily on physical measurements of *bathymetry* (seafloor depth) and ocean currents, but increasingly expeditions included naturalists. They collected specimens from the ocean as well as from land, along the shoreline and in deeper water using crude nets. Sailing on the HMS *Beagle* (1832–1836) inspired Darwin's revolutionary theory of evolution, but as the *Beagle* sailed from Africa to South America to Australia, Darwin also made important collections and observations on marine species such as corals.

The scientific leader of another influential expedition, Charles Wyville Thompson, on the HMS *Challenger* (1872–1876) had already led several deep-sea expeditions that proved the existence of life in the deep. The *Challenger* voyage was truly global, traversing the Atlantic, Pacific, and Southern Oceans. The diversity of life in a wide range of ocean habitats reported in the 50 volumes and 29,552 pages of the *Challenger Reports* comprise a scientific achievement not since repeated. At a cost of well over 10 million pounds in today's currency, it was truly a remarkable investment for its time. Indeed, the comprehensive and coordinated sampling detailed in the *Challenger Reports* stands as a model for the 10 years of exploration of the Census.

But the twentieth century saw the disappearance of large multiyear expeditions on the scale of the *Challenger* route. The cost of mounting such expeditions, and the availability and willingness of scientists to disappear for years, has effectively eliminated that model. That the examples of Darwin, seasick for the early stages of the *Beagle* voyage and Wyville Thompson's death at age 52, exhausted by leading and reporting on the *Challenger* expedition, are not lost on those of us who study the oceans! Importantly, the naturalists who led research activities in the 1800s had obligations to a specific employer for collecting and reporting on the material from the expedition. Today, however, ocean explorers work for universities and governments and have research and teaching commitments that preclude voyages of months or years. Further, few governments are willing to invest tens of millions of dollars in a single voyage around the world! Thus, oceanographic voyages, euphemistically called "cruises" today, are rarely longer than days to a month or two, rather than the 41 months of the *Challenger* expedition.

Exploration of marine biodiversity today

Although grand expeditions like the *Challenger* have ceased, today marine ecologists have advantages over those who studied the ocean a century ago. With technologies like satellites, underwater cameras, submersibles, and a whole range of sensors, modern explorers can see the ocean better than just by working over the side of a ship (Figure 3.1). Telephones, computers, and data transfer, often wireless, speed communication and create the potential for international collaboration like the Census, which would have taken many years to coordinate in the past. Communication with and data transfer from scientific instruments and even diving animals like sea lions far away is becoming routine. The *Challenger* expedition reported in dusty published tomes stored in the recesses of libraries scattered all over the world. But everyone can access the Census discoveries at the *Ocean Biogeographic Information System* (OBIS, www.iobis.org), an open-access, globally distributed network of taxonomic, environmental, and geographical information, with a double click online.

OBIS and other Internet sites have coordinated taxonomy and developed identification tools. Unlike printed identification guides and keys in books,

Figure 3.1 Sampling Planet Ocean.
To explore and sample Planet Ocean, scientists lowered bottles, nets, and grabs over the rails of ships to collect water and sediments, or bring back specimens. Later, scientists visited the deep ocean in person or launched vehicles they could drive from afar to collect samples.
(a) The recovery of a conductivity–temperature–depth (CTD) sampling rosette from the Beaufort Sea to study waters of the western Arctic. Niskin bottles surrounding the CTD close at defined depths to collect water for nutrient and microbial analysis aboard ships. See Chapter 7.
(b) A box corer lowered over the rail of a ship samples sediments on the seafloor beneath coastal and deep-sea waters. See Chapters 5 and 8. The metal spade on the left side of the sampler swings down after the metal box penetrates the seafloor, sealing the sample inside for transport to the surface.
(c) A pair of bongo nets towed behind a ship to collect two samples at once of zooplankton and larval fish. See Chapter 7.

(d)

(e)

(f)

(d) Three explorers dive in *Alvin*, the world's best-known deep-ocean submersible, which descends as deep as 4,500 meters (almost 3 miles), accessing two-thirds of the global seafloor. *Alvin* has explored the deep seafloor including hydrothermal vents (Chapter 8), photographed the sunken *Titanic*, and helped recover a lost hydrogen bomb.

(e) ROPOS, a Canadian remotely operated vehicle (ROV), sends video along a tether to pilots and scientists aboard a surface ship as the ROV samples down to 2,500 meters.

(f) Testing REMUS 6000, an autonomous underwater vehicle (AUV) at the dock in Woods Hole, Massachusetts. AUVs are untethered and sent off on programmed missions, such as mapping hydrothermal vents. See Chapter 8. REMUS recently searched the deep Atlantic for wreckage of an Air France flight.

Internet guides can be updated quickly and inexpensively, and users instantly notified. Digitized information on the Internet linked through the Census partner project *Encyclopedia of Life* (EOL, www.eol.org) has removed the need to track down obscure monographs or journals in distant libraries. Ecologist E. O. Wilson, who imagined "an electronic page for each species of organism on Earth, available everywhere by single access on command," inspired EOL. Its "species pages," once completed by specialists on that species, will provide instant links to published and unpublished material. EOL exemplifies how electronic communication accelerates the assembly, communication, and revision of knowledge and thus the movement of unknown to the known.

Up-to-date information like ever-improving species pages and taxonomic keys avoids mistaken identities that might take decades to resolve. Worse, the stain of unresolved mistakes propagates, even in the published literature, spreading quickly, while the origins of the error fade. Internet tools help marine ecologists who are not specialized taxonomists provide accurate knowledge of species resting on the soundest information about who's who in the ocean. Because species Web pages can link to related articles, efficiency improves because students of a species do not have to start from scratch assembling available information from widely scattered sources. Keep in mind the addition of something like 1,635 new marine species every year, and a single marine sample can contain dozens or even hundreds of unique, but easily confused species, initiating errors that quickly spread. Modern computers help stay on top of the burgeoning, mind-boggling information that once arrived in books, printed journals, and snail mail.

New molecular techniques differentiate among species, even look-alike species, and accelerate discovery. Because many microbes look alike, molecular techniques have opened up a previously unknown world of microbial diversity. Indeed, the opinion that for microbes "everything is everywhere," is now questioned. Individual species or different strains of a single microbial species can look identical, but have profoundly different impacts. For example, the toxic forms of the dinoflagellate (a single-celled protist) that causes red tides and paralytic shellfish poisoning are indistinguishable from benign strains, except by genetic analyses. These techniques also help evaluate evolutionary relationships among species, and even among places.

The Census differentiates species with barcoding technology, a unique identifier like the barcodes at grocery stores. Barcodes, using a segment of the cytochrome oxidase I (COI) gene from a cell's power pack, its mitochondria, differentiate most species of animals, both on land and in the water. Barcodes can rapidly tell us whether two specimens are the same species with relatively little ambiguity. When a library of barcodes has accumulated, any specimen can be identified by sequencing this gene and matching its barcode to one in the library. The Census partner project *Marine Barcode of Life* (MarBOL, www.marinebarcoding.org) is compiling a marine library and expects to accumulate reference codes for 50,000 species by the end of 2010. Barcoding will quickly and reliably determine whether any newly collected specimens have already been cataloged. Barcodes will minimize past confusion that regional variability causes in the appearance of some species and similarity in the appearance of others. For example, the shape of exposed kelps differs from the shapes of sheltered ones, but transplanting to a contrasting environment causes them to grow in a different shape adapted to the new flow conditions, something known as *phenotypic plasticity*. Barcoding tells us they are the same species, despite variation in form. Barcodes and their library will identify samples collected from unexplored places that have no taxonomic keys or experts. Barcodes and their library will help draw global maps of species distribution.

By avoiding mistaken identity, barcoding can help avoid confused management. When high school students collaborating with the Census sampled fish markets and restaurants in New York City, their barcoding found cheap, farmed fish labeled as expensive wild species, and an endangered species sold as a common one.

Contaminants in blue mussels are used to monitor coastal environments around the world. But three blue mussel species look alike and their distributions overlap, but have different growth rates and physiologies. Similarly, *Montastraea annularis*, the most widespread and studied coral in Caribbean reefs, encompasses sibling species that look alike, but grow at different rates. Confusion of species with different numbers of offspring, age at maturity, and growth confuses management because these characteristics dramatically change how quickly populations break down contaminants, or grow and replace themselves.

The recent rapid advance in the *Geographic Information System* (GIS) maps where organisms live. GIS speeds biodiversity studies, first by maps and then by visualization described below. GIS puts biological samples at the correct latitude and longitude, where overlaid maps of depth and temperature help us understand why species live where they do. Comprehensive maps of

life on the abyssal plains, at hydrothermal vents, or at the poles hinge
on such tools. Although GIS technology works best in two dimensions,
advances promise that the three dimensions of the ocean will soon become
visible. For example, Google Earth already displays Census observations
in their Ocean layer.

Where are the hot spots of diversity, where are fishing impacts pro-
nounced, is climate change redistributing species, and where and why do
animals migrate? GIS helps draw overlays of environmental variables on
maps to test predictions about the changing distribution and abundance
of ocean life.

Internet databases, including the Census' OBIS, are built on accurate
latitude and longitude. Formerly, explorers sampled at sea and published
their observations. Integrating such observations into a broader geographic
perspective assumed similar quality of identification and uses of species names
by others, and that they would share their data. This left much data on file
cards, yellowing datasheets in file cabinets, or on computer hard drives
around the world, which are often lost as experts retire.

Accordingly, "data rescue" avoids repetitive research and builds
multiple datasets into larger, more comprehensive analyses, all without
additional months of sailing and steaming. These *metadata* analyses, using
data from multiple sources, look at global or other broad patterns because few
individuals can collect enough data for broad comparisons across space
or time. With long stretches and vast depth, Planet Ocean frustrates lone
researchers. The Census repository OBIS therefore amalgamates separate
datasets from around the world to archive the lasting legacy of marine
biodiversity for analysis on regional to global scales. Sound metadata
analyses need the correct GIS locations and verified names in OBIS.

Bringing together experts from many countries, shores, and seas,
along with their ships and tools, the Census coordinated a global series of
interlinked expeditions. Linked projects developed genetic barcoding tools,
mapping techniques, global databases, and assembled samples for the
world's first global Census of Marine Life.

How many species live in the ocean?

Censuses, beginning with the Domesday Book collected for William the
Conqueror, count the single species of humans and tell us which are barons,

villains, or priests, and who has fish ponds and hides. In this spirit, the question of how many species live in the ocean is at the core of the Census of Marine Life. Even after a decade of intense research built on centuries of exploration, however, a precise answer is still elusive. The Nobel Prize winner Ernest Rutherford stated that, "All science is either physics or stamp collecting." Some regard biodiversity studies as a futile "hobby" of stamp collecting, not too different from compiling lifetime lists of birds spotted.

Recently, however, reports of species extinctions, and hence lost biodiversity along with ecological services, have made biodiversity knowledge valuable to Planet Ocean's top carnivores, humans. Sympathy grows for losing species before they are known to exist. Finally, most people recognize the value of knowledge, irrespective of economic payoff or impact on ecosystems.

Understanding what lives in the ocean begins with answering how many marine species scientists have named. How many marine species are known so far? Several recent analyses suggest somewhere between one-quarter and one-third of a million. A less ambitious, but necessary, first step is compiling global lists of species for a few major groups such as fishes. Most group lists are regional ones that overlap other lists and do not always identify organisms accurately. Is a specimen called species X from one location the same as one called X elsewhere? Or, are species called X and Y in different locations actually different? There are numerous examples of closely related *sibling species* that look alike, but have distinct biology and genetics.

Unaware what others have done recently, errors are made. Because scientists discover 1,300–1,500 valid new marine species every year, the reference list of known species constantly shifts. The shifts require frequent updates of keys and guides and render paper publications out of date as soon as they come off the press.

Fortunately, e-mail runs faster. Tracey Sutton, a Census fish specialist collecting in the South Atlantic, could hardly believe his luck when he looked at the animals in a trawl and realized that one of the juvenile fishes was intermediate between two families of fishes (Figure 3.2). E-mailing a photograph of the fish to an Australian colleague, he learned that the oddity was the final missing link to a puzzle they were putting together with a collaborator at the Smithsonian Institution. Genetic confirmation proved that three "different" groups of fishes with very different anatomies, previously thought to be totally different taxonomic families, are

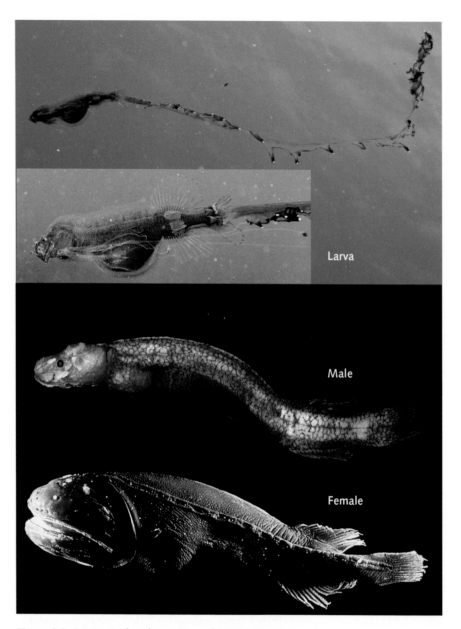

Larva

Male

Female

Figure 3.2 Anatomical and genetic analysis, sharp eyes, and international collaboration solve a long-standing conundrum.
After scientists had classified each of these three fishes as different families, more detailed morphological analysis and new specimens revealed that they are actually larvae, males, and females of a single family of whalefishes, the Cetomimidae. This revelation illustrates how even relatively well-studied fishes continue to yield exciting surprises.

actually the male, female, and larval stages of closely related fishes
from a single family.

Photographs can confuse as well as resolve. Fuzzy photographs
of a worm-like deep-sea animal assigned to the phylum Hemichordata,
led biologists to think it represented a missing evolutionary link between
two groups. But when a specimen of the since-named *Torquarator bullocki*
replaced the fuzzy photograph, the mystery was solved. Though not the
expected missing link, it was a new species, a new genus, and a new family.

Taxonomists may specialize in identifying a group as broad as all
polychaete worms or, more frequently, they specialize in a smaller subset.
The number of trained taxonomists is few, and is declining as museums cut
budgets and retiring experts are not replaced. Dedicated amateur hobbyists
augment these professionals incompletely and unevenly, specializing in just
a few groups, such as shelled mollusks and crustaceans, and leaving other
groups to the shrinking handful of professionals scattered around the world.
Because amateur taxonomists mainly see land species, between 1998 and
2005 they described only 10 to 15% of new European marine species, far
less than the 46% of new land and freshwater insects they described. This
neglect of the oceans contributes to the naming of new species in the ocean
lagging far behind naming on land, despite great diversity in the ocean.
Further, the location of most taxonomic expertise in Europe and North
America does not match the location of the bonanza of new species
found and to be discovered near tropical, developing countries.

Most specimens are collected by ecologists in their net tows and
bottom grabs, and they try to identify everything they catch. This scope
means they must identify species from many taxonomic groups. Although
their knowledge may be broad, they are not attuned to spotting new species
that are easily confused with known species. Simple mistakes in spelling or
taxonomists changing species names can further confuse ecologists' lists.
The good fortune of working with a taxonomist improves ecologists'
chances, but they rarely have expert partners for the multiple groups
they encounter.

The *World Register of Marine Species*, or WoRMS (www.marinespecies.org),
works with the Census, EOL, and OBIS, and the global list of species rapidly
improves. The organizations expect to confirm 250,000 valid, distinct species
by the fall of 2010 when this book is published. This goal can be achieved
because experts on many organisms worked together, carefully examining each
species to be certain it is unique and valid.

Putting aside the problem of misnamed species that the collaborating organizations hope to solve, I return to the still unanswered question, "How many *unknown* species live in the vast, thinly sampled ocean?" A wise answer recognizes the biases that affect the count as descriptions of new species in the ocean proceed and discovery moves the unknown to known. First, census takers count big and ostentatious organisms better than small, hidden microbes and invertebrates. Second, tropical oceanic islands, polar waters, seamounts, and the deep sea far from developed coastlines are more thinly explored than accessible areas like the European coastline. The tricky problem of estimating numbers of unknown species has inspired several strategies, each with limitations.

Extrapolating the rate at which new species are now being discovered can project future discoveries in groups where most species are known. For example, if new species are rarely added, as is now true for marine mammals, then it is fairly safe to conclude that future exploration will add few. But both accelerating and even continuing rates of discovery make projections difficult. In Figure 3.3, the curved thick line on the plot represents one geographic region or one group of organisms that is beginning to level off, making 100 a reasonable estimate of the total number of species. But if the curve for a region or group has not begun to plateau, like the straight thick line, then there is no way to know how much higher it will climb before it eventually levels off. Its total could be 500, 1000, or many more. Also the slopes of the curves will rise faster or slower depending on how many people are describing new species. Projecting the number of species and thus diversity in the ocean suffers from the timeless conundrum: when will a trend end?

Fishes provide a specific example of numbers. A thorough global list (www.fishbase.org) resting on the widespread counting of fish makes possible a precise estimate of 16,475 species. The evident rate of discovery and its deceleration enable extrapolation to the eventual total. The Census project, the *Future of Marine Animal Populations* (FMAP), did exactly this for fish, where they estimated that 21% of all fishes in the ocean have yet to be discovered. They estimate between 1,000 and 4,000 more species to be found, depending on the statistical model. Not surprisingly, because some environments are less known than others, the deep-sea *bathyal* (1,000 to 4,000 meters) and *demersal* (seafloor) habitats as well as the tropical environments are yielding and will continue to yield most new species

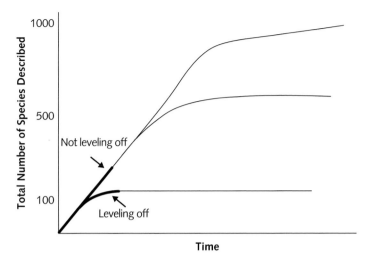

Figure 3.3 How many more species remain to be found?
Estimating unknown marine diversity is a tricky business. A theoretical curve
illustrates that if nothing changes as time passes, more species are discovered and
described, year by year. This theory applies to one group or species or to all the species in
one environment. If the curve begins to level off (lower curve), then statistical models
can estimate when the number of species will stop rising, indicating that all the species in
a given group or given environment have been discovered. But if the curve shows no
signs of leveling off (middle and upper curves), then discovery will continue, making
it impossible to know which of those two curves will eventually form as knowledge
of new species accumulates.

of fishes. Even this relatively certain estimate for a well-known group
rests on incomplete data and erroneous names. Also, scientists believe that
Caribbean fishes are actually not well explored and may yield many more
species than predicted with this approach.

The expert opinion of taxonomic specialists for specific groups
provides an alternative. An expert on marine mollusks who has studied
specimens from a variety of environments could estimate how many new
species are typically found when new environments are explored and then
extrapolate globally. If experts from groups of organisms are asked for
estimates for their group, pooling those estimates together or scaling
upward from well-known groups to all groups can build an estimate of
the total number of species in the ocean. This approach estimates 500,000
to 1,000,000 multicellular species. Still another approach adds ocean habitats
geographically. Because many experts focus on one type of habitat, such as
coral reefs, or particular groups of organisms that are more common in

some habitats than others, their estimate of unknowns for that habitat
or group is more accurate. This approach estimates that, not including
microbes, coral reefs alone may be home to 600,000 to 9.5 million species.

Asking how many new species will be found if more ocean is
explored forms still another estimator. Exhaustive sampling presumably
finds all species in a small lake, for example, or reaches the point where
the unknown species can be estimated with reasonable precision. Fred
Grassle and Nancy Maciolek performed this analysis for the deep sea in 1992.
They estimated the number of *macrofaunal* species, the invertebrates such as
polychaete worms and mollusks from 0.3 millimeters to several centimeters in
size, regionally and globally. They compiled the most complete dataset of its
time for all macrofauna from any region of the deep ocean. They sampled
primarily along a depth contour at 2,100 meters, adding species with
each new sample and distance along the contour.

Grassle and Maciolek reasoned that rates of adding species along a
constant depth should conservatively estimate how more exploration will
lift the total of new species. Extrapolating from their area of exploration
along the 2,100-meter contour out to the rest of the deep sea, they estimated
100 million species. Because the abyssal plains have few macrofauna, they
scaled back their estimate by an order of magnitude to 10 million species.

This estimate prompted debate whether 10 million was too high or
whether including smaller *meiofauna* such as nematode threadworms and
tiny crustaceans (less than 0.3 millimeters, but larger than 0.04 millimeters),
would raise the estimate. A critical question is which species are *cosmopolitan*,
or widely distributed geographically. Rapidly changing composition with
geographic distance would raise Grassle and Maciolek's estimate, whereas
slow change would lower it. One specialist who estimated 100s of millions
of deep-sea meiofaunal species later downgraded the estimate to less
than one million, since many are cosmopolitan.

Others proposed estimating unknown marine species numbers
by the percentage of new species found in each study. A criticism of
the Grassle and Maciolek estimate was that they found only 58% of their
species were new to science. Extrapolating that percentage by the number
of described species in the ocean suggests that the deep sea contains perhaps
250,000 macrofaunal species. The major caveat with this use of published
studies is that not all areas have been studied equally, and in the deep South
Pacific the proportion of new species was as much as 95%, and suggested
an extrapolation of 500,000 macrofaunal species was more appropriate.

Because 10 million is the estimated number of species in the deep ocean, answering how many species live in the entire ocean brings a larger number. Exploration of places other than the deep sea also adds to the total of marine species. For example, 20% of European gastropods (snails and related organisms) were described within the last 25 years, which extrapolates to 11–50% of European marine fauna remaining to be found. Mollusk expert Philippe Bouchet found more mollusk species in small samples from a single 30,000 hectare area of the tropical Pacific than are known for all of the Mediterranean, which is 10,000 times larger!

Encompassing the wild card of microbes in the count of species will also add to the number of species in the ocean greatly. Whether species are widely distributed is least known for smaller organisms. Microbial experts argue that a single drop of seawater may hold 160 kinds of bacteria, and in the deep sea a liter of seawater may hold thousands of different kinds of microbes, including rare forms represented by a few or even one individual. Protistologists argue whether protist distributions are cosmopolitan, but the few sparsely distributed samples of such organisms cannot possibly resolve the debate. Given that little of the ocean has been sampled for microbial diversity, how is it possible to scale up samples from a few liters of water or sediment to the entire ocean? At this point it isn't, at least without more sampling. By analogy, imagine trying to estimate the number of species on an entire continent based only on observations from your own backyard!

While Bouchet estimated that known marine biodiversity represents about 15% of all species on Earth, he also noted slower discovery in marine than terrestrial fauna. He calculated that explorers will spend 250 years to a millennium to finish inventorying marine biodiversity, ignoring microbes, hindered by limited exploration and taxonomists. In recent years, some countries have prohibited foreigners from sampling in their territorial waters because biodiversity may have unknown economic potential. Because poorer countries lack the scientific capacity to explore their own waters, the global ocean may not be explored for a long time. Still, the emerging tools described above will help to shorten the time, but the time will remain long and will depend on how quickly poor nations can develop scientific capability.

To accelerate the taxonomic tide to Bay of Fundy speed, the *Census of Marine Zooplankton* (CMarZ) sponsored some 27 taxonomy workshops involving 253 taxonomists who marry traditional morphological taxonomy with barcoding. Online resources and imagery such as the EOL, in tandem with accurate inventories such as WoRMS and online keys and databases will also make taxonomy easier and more precise. These efforts will shorten century-long projections. Moreover, the excitement from Census discoveries such as the Yeti crab (Figure 1.5a) and other species found near and far during the last decade will inspire marine exploration.

Even a large, international program like the Census of Marine Life pushes the frontier of the known forward slowly. Although experts continue to argue about the magnitude – even the order of magnitude – of marine biodiversity, the numbers rest on more facts. Ignoring the wildcard of microbes, most scientists would probably be comfortable saying that there are at least one million species in the ocean, or maybe several million. Accuracy, or closeness to the real value, depends on knowing the level of uncertainty in those predictions, which new statistical models improve. The new technology examined in the next chapter will expand the span of exploration, making extrapolations more accurate and defensible.

BIBLIOGRAPHY

Adrianov, A. V., C. Murakami, and Y. Shirayama, Taxonomic study of the kinorhyncha in Japan. III. *Echinoderes sensibilis n. sp* (Kinorhyncha: Cyclorhagida) from Tanabe Bay. *Zool. Sci.*, **19**:4 (2002), 463–73.

Agostini, V., A. S. Arico, E. E. Briones, *et al.*, *Global Open Oceans and Deep Seabed (GOODS) – Biogeographic Classification* (Paris: UNESCO-IOC, 2009).

Amaral-Zettler, L., L. F. Artigas, J. Baross, *et al.*, A global census of marine microbes. In A. D. McIntyre (ed.), *Life in the World's Oceans: Diversity, Distribution, and Abundance* (Oxford: Blackwell Publishing Ltd., 2010), pp. 223–45.

Baker, M. C., E. Z. Ramirez-Llodra, P. A. Tyler, *et al.*, Biogeography, ecology and vulnerability of chemosynthetic ecosystems in the deep sea. In A. D. McIntyre (ed.), *Life in the World's Oceans: Diversity, Distribution, and Abundance* (Oxford: Blackwell Publishing Ltd., 2010), pp. 161–82.

Baum, J. K. and R. A. Myers, Shifting baselines and the decline of pelagic sharks in the Gulf of Mexico. *Ecol. Lett.*, **7**:2 (2004), 135–45.

Baum, J. K., R. A. Myers, D. G. Kehler, *et al.*, Collapse and conservation of shark populations in the Northwest Atlantic. *Science*, **299**:5605 (2003), 389–92.

Bekker-Nielsen, T. (ed.), *Ancient Fishing and Fish Processing in the Black Sea Region* (Oxford: Aarhus University Press, 2006).

Billett, D. S. M., B. J. Bett, W. D. K. Reid, B. Boorman, and I. G. Priede, Long-term change in the abyssal NE Atlantic: the "*Amperima* Event" revisited. *Deep Sea Res. II*, **50**:1–4 (2009), 325–48.

Block, B. A., H. Dewar, S. B. Blackwell, *et al.*, Migratory movements, depth preferences, and thermal biology of Atlantic bluefin tuna. *Science*, **293**:5533 (2001), 1310–4.

Boero, F. and C. E. Mills, Hydrozoan people come together. *Trends Ecol. Evol.*, **14**:4 (1999), 127–8.

Bouchet, P., The magnitude of marine biodiversity. In C. M. Duarte (ed.), *The Exploration of Marine Biodiversity: Scientific and Technological Challenges* (Madrid, Spain: Fundacion BBVA, 2006).

Bousfield, E. L., *Shallow-water Gammaridean Amphipoda of New England* (Ithaca, New York: Cornell University Press, 1973).

Bowman, J. P., C. Mancuso, C. M. Nichols, and J. A. E. Gibson, *Algoriphagus ratkowskyi* gen. nov., sp nov., *Brumimicrobium glaciale* gen. nov., sp nov., *Cryomorpha ignava* gen. nov., sp nov. and *Crocinitomix catalasitica* gen. nov., sp nov., novel flavobacteria isolated from various polar habitats. *Int. J. Syst. Evol. Microbiol.*, **53** (2003), 1343–55.

Branch, G. M., C. L. Griffiths, M. L. Branch, and L. E. Beckley, *Two Oceans: A Guide to the Marine Life of Southern Africa* (South Africa: Struik Publishers, 1994).

Brandt, A., New species of Nannoniscidae (Crustacea, Isopoda) and *Saetoniscus* n. gen. from the deep sea of the Angola Basin. *Zootaxa*, **88** (2002), 1–36.

Brandt, A., C. De Broyer, I. De Mesel, *et al.*, The biodiversity of the deep Southern Ocean benthos. *Philos. Trans. R. Soc. Lond. B Biol. Sci.*, **362**:1477 (2007), 39–66.

Bouchet, P., Inventorying the molluscan diversity of the world: what is our rate of progress? *The Veliger*, **40**:1 (1997), 1–11.

Bouchet, P., P. Lozouet, P. Maestrati, and V. Heros, Assessing the magnitude of species richness in tropical marine environments: exceptionally high numbers of molluscs at a New Caledonia site. *Biol. J. Linn. Soc.*, **75**:4 (2002), 421–36.

Briggs, J. C., Species diversity: land and sea compared. *Syst. Biol.*, **43**:1 (1994), 130–5.

Brokeland, W., Three species of the isopod crustacean genus *Antennuloniscus menzies*, 1962 (Asellota: Haploniscidae) from the Southern Ocean. *Zootaxa*, **1115** (2006), 1–29.

Brokeland, W. and A. Brandt, Two new species of Ischnomesidae (Crustacea: Isopoda) from the Southern Ocean displaying neoteny. *Deep Sea Res. II*, **51**:14–16 (2004), 1769–85.

Bucklin, A., S. Nishida, S. Schnack-Schiel, *et al.*, A census of zooplankton of the global ocean. In A. D. McIntyre (ed.), *Life in the World's Oceans: Diversity, Distribution, and Abundance* (Oxford: Blackwell Publishing Ltd., 2010) pp. 247–65.

Chao, L. S. L., R. E. Davis, and C. L. Moyer, Characterization of bacterial community structure in vestimentiferan tubeworm *Ridgeia piscesae* trophosomes. *Mar. Ecol.*, **28**:1 (2007), 72–85.

Chernova, N. V. and P. R. Moller, A new snailfish, *Paraliparis nigellus* sp nov (Scorpaeniformes, Liparidae), from the Northern Mid-Atlantic Ridge – with notes on occurrence of *Psednos* in the area. *Mar. Biol. Res.*, **4**:5 (2008), 369–75.

Chisholm, S. W., R. J. Olson, E. R. Zettler, *et al.*, A novel free-living prochlorophyte abundant in the oceanic euphotic zone. *Nature*, **334**:6180 (1988), 340–3.

Consalvey, M., M. R. Clark, A. A. Rowden, and K. I. Stocks, Life on seamounts. In A. D. McIntyre (ed.), *Life in the World's Oceans: Diversity, Distribution, and Abundance* (Oxford: Blackwell Publishing Ltd., 2010) pp. 123–38

Cristobo, F. J., V. Urgorri, and P. Ríos, Three new species of carnivorous deep-sea sponges from the DIVA-1 expedition in the Angola Basin (South Atlantic). *Org. Divers Evol.*, **5**:Suppl. 1 (2005), 203–13.

Crosnier, A. and A. Vereshchaka, *Altelatipes falkenhaugae* n. gen., n. sp. (Crustacea, Decapoda, benthesicymidae) de la Ride Médio-Atlantique Nord. *Zoosystema*, **20**:2 (2008), 399–411.

Curtis, T. P., W. T. Sloan, and J. W. Scannell, Estimating prokaryotic diversity and its limits. *Proc. Natl. Acad. Sci. USA*, **99**:16 (2002), 10494–9.

Darwin, C., *The Zoology of the Voyage of HMS Beagle* (London: Smith Elder and Co., 1839).

Díaz, J. M., F. Gast, and D. C. Torres, Rediscovery of a Caribbean living fossil: *Pholadomya candida* GB Sowerby I, 1823 (Bivalvia: Anomalodesmata: Pholadomyoidea). *Nautilus*, **123**:1 (2009), 19–20.

Ebbe, B., D. Billett, A. Brandt, *et al.*, Diversity of abyssal marine life. In A. D. McIntyre (ed.), *Life in the World's Oceans: Diversity, Distribution, and Abundance* (Oxford: Blackwell Publishing Ltd., 2010), pp. 139–60.

Erwin, T. L., Tropical forests: their richness in coleoptera and other arthropod species. *Coleopterists Bull.*, **36** (1982), 74–5.

Eschmeyer, W. N., R. Fricke, J. D. Fong, and D. Polack, Marine fish biodiversity: history of knowledge and discovery (Pisces). *Zootaxa* **2525**(2010), 19–50.

Finlay, B. J. and T. Fenchel, Cosmopolitan metapopulations of free-living microbial eukaryotes. *Protist*, **155**:2 (2004), 237–44.

Froese, R. and D. Pauly, *FishBase*. 2009 (cited 2010). Available from: www.fishbase.org.

Fukami, H., A. F. Budd, G. Paulay, *et al.*, Conventional taxonomy obscures deep divergence between Pacific and Atlantic corals. *Nature*, **427**:6977 (2004), 832–5.

Gagaev, S. Y., *Sigambra healyae* sp n., a new species of polychaete (Polychaeta: Pilargidae) from the Canadian Basin of the Arctic Ocean. *Russ. J. Mar. Biol.*, **34**:1 (2008), 73–5.

Gage, J., Evolution – deep-sea spiral fantasies. *Nature*, **434**:7031 (2005), 283–4.

Gerken, S., Two new *Cumella* (Crustacea: Cumacea: Nannastacidae) from the North Pacific, with a key to the North Pacific Cumella. *Zootaxa*, **2149** (2009), 50–61.

Glover, A. G., E. Goetzel, T. G. Dahlgren, and C. R. Smith, Morphology, reproductive biology and genetic structure of the whale-fall and hydrothermal vent specialist, *Bathykurila guaymasensis* Pettibone, 1989 (Annelida: Polynoidae). *Mar. Ecol.*, **26**:3–4 (2005), 223–34.

Glover, A. G., B. Kallstrom, C. R. Smith, and T. G. Dahlgren, World-wide whale worms? A new species of *Osedax* from the shallow North Atlantic. *Proc. R. Soc. Lond. B Biol. Sci.*, **272**:1581 (2005), 2587–92.

Gradinger, R., B. A. Bluhm, R. R. Hopcroft, *et al.*, Marine life in the Arctic. In A. D. McIntyre (ed.), *Life in the World's Oceans: Diversity, Distribution, and Abundance* (Oxford: Blackwell Publishing Ltd., 2010), pp. 183–202.

Grassle, J. F. and J. P. Grassle, Opportunistic life histories and genetic systems in marine benthic polychaetes. *J. Mar. Res.*, **32**:2 (1974), 253–84.

Grassle, J. P. and J. F. Grassle, Sibling species in the marine pollution indicator *Capitella* (polychaeta). *Science*, **192**:4239 (1976), 567–9.

Grassle, J. F. and N. J. Maciolek, Deep-sea species richness – regional and local diversity estimates from quantitative bottom samples. *Am. Nat.*, **139**:2 (1992), 313–41.

Grassle, J. F. and K. I. Stocks, A global Ocean Biogeographic Information System (OBIS) for the Census of Marine Life. *Oceanography*, **12** (1999), 12–4.

Greene, C. H., A. J. Pershing, T. M. Cronin, and N. Ceci, Arctic climate change and its impacts on the ecology of the North Atlantic. *Ecology*, **89**:11 (2008), S24–S38.

Griffiths, H. J., B. Danis, and A. Clarke, Quantifying Antarctic marine biodiversity: the SCAR-MarBIN data portal. *Deep Sea Res. II* (2010).

Groeneveld, J. C., C. L. Griffiths, and A. P. Van Dalsen, A new species of spiny lobster, *Palinurus barbarae* (Decapoda, Palinuridae) from Walters Shoals on the Madagascar Ridge. *Crustaceana*, **79** (2006), 821–33.

Groombridge, B. and M. D. Jenkins, *World Atlas of Biodiversity: Earth's Living Resources in the 21st Century* (California: University of California Press, 2002).

Gutt, J., G. Hosie, and M. Stoddart, Marine life in the Antarctic. In A. D. McIntyre (ed.), *Life in the World's Oceans: Diversity, Distribution, and Abundance* (Oxford: Blackwell Publishing Ltd., 2010), pp. 203–20.

Hartman, O., *Atlas of Sedentariate Polychaetous Annelids from California* (Los Angeles: University of Southern California, 1969).

Hebert, P. D. N., A. Cywinska, S. L. Ball, and J. R. DeWaard, Biological identifications through DNA barcodes. *Proc. R. Soc. Lond. B Biol. Sci.*, **270**:1512 (2003), 313–21.

Holland, N. D., D. A. Clague, D. P. Gordon, *et al.*, 'Lophenteropneust' hypothesis refuted by collection and photos of new deep-sea hemichordates. *Nature*, **434**:7031 (2005), 374–6.

Holm, P., A. H. Marboe, B. Poulsen, and B. R. MacKenzie, Marine animal populations: a new look back in time. In A. D. McIntyre (ed.) *Life in the World's Oceans: Diversity, Distribution, and Abundance* (Oxford: Blackwell Publishing Ltd., 2010), pp. 3–23.

Jackson, J. B. C., M. X. Kirby, W. H. Berger, *et al.*, Historical overfishing and the recent collapse of coastal ecosystems. *Science*, **293**:5530 (2001), 629–38.

Johnson, G. D., J. R. Paxton, T. T. Sutton, *et al.*, Deep-sea mystery solved: astonishing larval transformations and extreme sexual dimorphism unite three fish families. *Biol. Lett.*, **5**:2 (2009), 235–9.

Jonsen, I. D., R. A. Myers, and M. C. James, Robust hierarchical state-space models reveal diel variation in travel rates of migrating leatherback turtles. *J. Anim. Ecol.*, **75**:5 (2006), 1046–57.

Knowlton, N., The future of coral reefs. *Proc. Natl. Acad. Sci. USA*, **98**:10 (2001), 5419–25.

Knowlton, N., R. E. Brainard, R. Fisher, *et al.*, Coral reef biodiversity. In A. D. McIntyre (ed.), *Life in the World's Oceans: Diversity, Distribution, and Abundance* (Oxford: Blackwell Publishing Ltd., 2010), pp. 65–77.

Knowlton, N., E. Weil, L. A. Weigt, and H. M. Guzman, Sibling species in *Montastraea annularis*, coral bleaching, and the coral climate record. *Science*, **255**:5042 (1992), 330–3.

Koehl, M. A. R., W. K. Silk, H. Liang, and L. Mahadevan, How kelp produce blade shapes suited to different flow regimes: a new wrinkle. *Integr. Comp. Biol.*, **48**:6 (2008), 834–51.

Komai, T. and M. Segonzac, A revision of the genus *Alvinocaris* Williams and Chace (Crustacea: Decapoda: Caridea: Alvinocarididae), with descriptions of a new genus and a new species of *Alvinocaris*. *J. Nat. Hist.*, **39**:15 (2005), 1111–75.

Koslow, J. A., A. Williams, and J. R. Paxton, How many demersal fish species in the deep sea? A test of a method to extrapolate from local to global diversity. *Biodivers. Conserv.*, **6**:11 (1997), 1523–32.

Kuklinski, P. and P. D. Taylor, A new genus and some cryptic species of Arctic and boreal calloporid cheilostome bryozoans. *J. Mar. Biol. Assoc. UK*, **86**:5 (2006), 1035–46.

Lambshead, P. J. D., Recent developments in marine benthic biodiversity research. *Oceanis*, **19**:6 (1993), 5–24.

Lambshead, P. J. D. and G. Boucher, Marine nematode deep-sea biodiversity – hyperdiverse or hype? *J. Biogeogr.*, **30**:4 (2003), 475–85.

Levin, L. A., D. F. Boesch, A. Covich, *et al.*, The function of marine critical transition zones and the importance of sediment biodiversity. *Ecosystems*, **4**:5 (2001), 430–51.

Lindsay, D., F. Pages, J. Corbera, *et al.*, The anthomedusan fauna of the Japan Trench: preliminary results from in situ surveys with manned and unmanned vehicles. *J. Mar. Biol. Assoc. UK*, **88**:8 (2008), 1519–39.

Linnaeus, C., *Systemae Naturae, sive regna tria naturae. Systematics proposita per Classes, Ordines, Genera & Species* (Lugduni Batavorum: Apud Theodorum Haak, 1735).

Linnaeus, C., *Systema Naturae per Regna Tria Naturae Secundum Classes, Ordines, Genera, Species, Cum Characteribus, Differentiis, Synonymis, Locis*, 12th edn (Stockholmiae: Laurentii Salvii, 1767).

Lopez-Gonzalez, P. J., J. M. Gili, and C. Orejas, A new primnoid genus (Anthozoa: Octocorallia) from the Southern Ocean. *Sci. Mar.*, **66**:4 (2002), 383–97.

Lutz, R. A., Deep-sea vents: science at the extreme. *Natl. Geogr. Mag.*, **198**:4 (2000), 116–27.

Macpherson, E., W. Jones, and M. Segonzac, A new squat lobster family of Galatheoidea (Crustacea, Decapoda, Anomura) from the hydrothermal vents of the Pacific-Antarctic Ridge. *Zoosystema*, **27**:4 (2005), 709–23.

Macpherson, E. and M. Segonzac, Species of the genus *Munidopsis* (Crustacea, Decapoda, Galatheidae) from the deep Atlantic Ocean, including cold-seep and hydrothermal vent areas. *Zootaxa*, **1095** (2005), 3–60.

Malyutina, M. and A. Brandt, *Gurjanopsis australis* gen. nov., sp nov., a new epibenthic deep-sea munnopsid (Crustacea, Isopoda, Munnopsidae) from the Weddell Sea, Southern Ocean. *Deep Sea Res. II*, **54** (2007), 1806–19.

May, R. M., Biodiversity – bottoms up for the oceans. *Nature*, **357**:6376 (1992), 278–9.

May, R. M., Biological diversity – differences between land and sea. *Philos. Trans. R. Soc. Lond. B Biol. Sci.*, **343**:1303 (1994), 105–11.

Mayr, E., *Systematics and the Origin of Species from the Viewpoint of a Zoologist* (New York: Columbia University Press, 1942).

Menot, L., M. Sibuet, R. S. Carney, *et al.*, New perceptions of continental margin biodiversity. In A. D. McIntyre (ed.), *Life in the World's Oceans: Diversity, Distribution, and Abundance* (Oxford: Blackwell Publishing Ltd., 2010), pp. 79–101.

Mora, C., D. P. Tittensor, and R. A. Myers, The completeness of taxonomic inventories for describing the global diversity and distribution of marine fishes. *Proc. R. Soc. Lond. B Biol. Sci.*, **275**:1631 (2008), 149–55.

Miloslavich P., J. M. Diaz, E. Klein, J. J. Alvarado, C, Diaz *et al.* Marine biodiversity in the Caribbean: Regional estimates and distribution patterns. *PLoS ONE* **5**:8 (2010), e11916, doi: 10.1371/journal.pone.0011916.

Murray, J., *Report of the Scientific Results of the Voyage of HMS Challenger during the year 1873–76* (1895).

McDonald, J. H., R. Seed, and R. K. Koehn, Allozymes and morphometric characters of 3 species of *Mytilus* in the northern and southern hemispheres. *Mar. Biol.*, **111**:3 (1991), 323–33.

PacificFishing, We scoop the New York Times. *Today's Fish Wrap* (2008). Available from: http://www.pacificfishing.com/archives/week_of_082508pf.html.

Poore, G. C. B. and G. D. F. Wilson, Marine species richness. *Nature*, **361**:6413 (1993), 597–8.

Reaka-Kudla, M. L., The global biodiversity of coral reefs: a comparison with rain forests. In M. L. Reaka-Kudla, D. E. Wilson, and E. O. Wilson (eds.), *Biodiversity II: Understanding and Protecting Our Biological Resources* (Washington, DC: Joseph Henry Press, 1997), pp. 83–108.

Rice, A. L., The Challenger Expedition: the end of an era or a new beginning? In M. Deacon, T. Rice, and C. Summerhayes (eds.), *Understanding the Oceans* (Abingdon, Oxon: Routledge, 2001), pp. 27–48.

Robertson, D. R., Global biogeographical data bases on marine fishes: caveat emptor. *Divers. Distrib.*, **14**:6 (2008), 891–2.

Rogacheva, A. V., Revision of the Arctic group of species of the family Elpidiidae (Elasipodida, Holothuroidea). *Mar. Biol. Res.*, **3**:6 (2007), 367–96.

Schindel, D. E. and S. E. Miller, DNA barcoding a useful tool for taxonomists. *Nature*, **435**:7038 (2005), 17.

Scholin, C., G. Doucette, S. Jensen, *et al.*, Remote detection of marine microbes, small invertebrates, harmful algae, and biotoxins using the environmental sample processor (ESP). *Oceanography*, **22**:2 (2009), 158–67.

Snelgrove, P. V. R., Getting to the bottom of marine biodiversity: sedimentary habitats. *BioScience*, **49** (1999), 129–38.

Snelgrove, P. V. R. and C. R. Smith, A riot of species in an environmental calm: the paradox of the species-rich deep-sea floor. *Oceanogr. Mar. Biol. Ann. Rev.*, **40** (2002), 311–42.

Sogin, M. L., H. G. Morrison, J. A. Huber, *et al.*, Microbial diversity in the deep sea and the underexplored "rare biosphere." *Proc. Natl. Acad. Sci. USA*, **103** (2006), 12115–20.

Soliman, Y. and M. Wicksten, *Ampelisca mississippiana*: a new species (Crustacea: Amphipoda: Gammaridea) from the Mississippi Canyon (Northern Gulf of Mexico). *Zootaxa*, **1389** (2007), 45–54.

Sutton, T. T. and K. E. Hartel, New species of *Eustomias* (Teleostei: Stomiidae) from the western North Atlantic, with a review of the subgenus *Neostomias*. *Copeia*, **1** (2004), 116–21.

Van Dover, C. L., C. R. German, K. G. Speer, L. M. Parson, and R. C. Vrijenhoek, Marine biology – evolution and biogeography of deep-sea vent and seep invertebrates. *Science*, **295**:5558 (2002), 1253–7.

Vecchione, M., L. Allcock, U. Piatkowski, and J. Strugnell, *Benthoctopus rigbyae*, n. sp., a new species of cephalopod (Octopoda; Incirrata) from near the Antarctic Peninsula. *Malacologia*, **51**:1 (2009), 13–28.

Ward, P. and R. A. Myers, Shifts in open-ocean fish communities coinciding with the commencement of commercial fishing. *Ecology*, **86**:4 (2005), 835–47.

Wilson, S. P. and M. J. Costello, Predicting future discoveries of European marine species by using a non-homogeneous renewal process. *J. R. Stat. Soc. Ser. C Appl. Stat.*, **54** (2005), 897–918.

Worm, B., E. B. Barbier, N. Beaumont, *et al.*, Impacts of biodiversity loss on ocean ecosystem services. *Science*, **314**:5800 (2006), 787–90.

Worm, B., M. Sandow, A. Oschlies, H. K. Lotze, and R. A. Myers, Global patterns of predator diversity in the open oceans. *Science*, **309**:5739 (2005), 1365–9.

Young, R. E., M. Vecchione, and U. Piatkowski, *Promachoteuthis sloani*, a new species of the squid family Promachoteuthidae (Mollusca: Cephalopoda). *Proc. Biol. Soc. Wash.*, **119**:2 (2006), 287–92.

PART TWO

The known: what has the Census learned?

4

New ways of seeing deeper and farther

For millenia, wading and sailing humans have collected ocean life to eat, and to cover and decorate themselves. They collected it for fuel and fertilizer. As burgeoning populations intensified harvesting and grew more dependent on marine life, their efforts to understand where organisms live and why they live there also grew. Fortunately, physics, geology, and chemistry advanced in parallel, improving technology for seeing deeper into the once opaque ocean, even to the ocean bottom.

The once opaque ocean

Google Earth shows the sunlit yard behind our homes, but sunlight penetrates only hundreds of meters at best through water. Sailors see little over the rail. The optical tools we lower into the ocean "see" only a few tens of meters through water before images become fuzzy and useless. New technologies and tools now peer farther into and across the ocean in ways that seem like Jules Verne's science fiction. To answer its questions, the Census invented some tools and adapted others. Other explorers independently developed tools. The new toolbox makes the once opaque ocean more transparent, clarifying the view deep below the surface.

Vast, fluid, and moving, the ocean runs far and deep. Swimming and drifting marine life travels among continents or inhabits hidden habitats far from shore, at great depths, or in very specialized habitats that are difficult to sample. People dive far less deeply into Planet Ocean, and see only a tiny fraction of the ocean before generalizing from their limited glimpses. It's like 60-second dating, where glimpses transform lives. Planet Ocean will not shrink – if anything, rising sea levels will expand it.

This means we must become smarter to see deeper and farther. Ships still carry us to different seas, where we lower buckets and nets into the depths beneath. More than a century ago, when HMS *Challenger* set to sea, researchers were sampling almost blindly because they knew little about what lay beneath. They were less afraid of sailing over the Earth's edge or monsters rising from the deep than Columbus' crew centuries earlier. Nevertheless, much of Planet Ocean still was a new frontier stretching where humans had never looked. The naming of about four new marine species daily testifies that, while twentieth-century explorers may be less blind to unknown ocean life than in the past, they still lack 20:20 vision.

Scientists often cruise separately to locations of their choosing, bringing knowledge together in hindsight rather than in planning. This scattered approach has created an incomplete "photo album" of Planet Ocean with many snapshots, but also many gaps. Scientists realize that merging datasets would produce a better "photo album" or even a "film" of all marine life. Mark Twain classified "lies, damned lies, and statistics," but experience has taught scientists that only censuses, numbers, and statistics measure whether Planet Ocean is changing, for better or worse.

The more transparent ocean

As scientists sample more completely, with better nets, remotely operated vehicles (ROVs), roaming autonomous underwater vehicles (AUVs), and manned submersibles (Figure 3.1), the ocean is becoming more transparent. Precision sampling of the water layers and seafloor now rivals what we can do on land. Exploring the sea, of course, costs more because sailing a research ship costs more than driving a pickup truck.

With sound waves from acoustic instruments, scientists count organisms from afar (Figure 4.1) and map bumps and mountains on the seafloor. Sensors mounted on AUVs "sniff out" chemical signatures from distant hydrothermal vents. Electronic tags on animals tell us where they are and measure their environment, before sending data to shore or

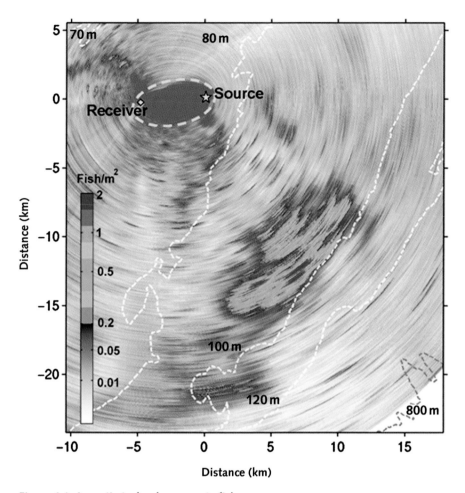

Figure 4.1 Acoustic technology counts fish.
Acoustic technology aboard a ship instantly imaged a school of fish the size of
Manhattan. Acoustic waves remotely sensed a shoal of more than 20 million fish,
almost certainly herring, off the coast of New Jersey. The white lines show depth
contours, and the colors show fish abundance. In about a minute, this new technology
images hundreds of cubic kilometers of ocean, expediting efforts to census how
many fish are in the sea. Sound waves sent from a source on one ship bounce
off of fish, which are counted by a receiver hanging over the side of a second
ship. See Chapter 5.

satellites. Via the Internet we can follow albatrosses as they leave their
feeding grounds in Alaska for their romantic Caribbean mating rendezvous.
Electronic tags can reveal whether "Tirion" the tuna (with apologies to
J. R. R. Tolkien) is dining near Tokyo or San Francisco this month.

New molecular tools solve taxonomic problems. Although one microbial cell may be indistinguishable from the next under highest magnification, a "genetic microscope" can differentiate the agent of lethal paralytic shellfish poisoning from a tasty mussel appetizer. Ann Bucklin, a leader of the *Census of Marine Zooplankton* (CMarZ) said, "We used to think we knew many species well, but the advent of DNA barcoding has radically altered that perception."

We've come a long way since Captain James Cook mapped the ocean with only a sextant, sun, and stars to tell him where he was, and dropped a long line weighted with lead and a little cup to measure depth and composition of the seafloor. GIS, *Geographic Information System*, has catalyzed new mapping techniques that show exactly where organisms live in the ocean, a first step is learning why they live there. Multibeam acoustics and echosounders tell us the depth and composition of the seafloor. Census researchers used mapping and other tools to follow migrating animals as they moved through the ocean and to link individuals and species to their environment. Census researchers even named turtles and invited the public to follow migrating individuals from day to day.

Living ocean: diversity, distribution, and abundance

The Census observed diversity, distribution, and abundance of ocean life. Although studies typically focus on just one of these attributes, the gears overlap. For longer than a century, fishermen and scientists alike collected ocean life with nets (Figure 3.1). They lowered bottles over the rail to collect small volumes of water and their microbes. They designed bottles early in the twentieth century and modified them in the 1960s with sterile plastic and Teflon coatings to avoid contamination. When declining catches and increasing curiosity drove fishermen and scientists into deeper water and more difficult habitats, they improved simple bags and scoops dating from ancient Greece.

Nets and bottle samplers collect some organisms well from some waters. Bottle samplers, however, collect small volumes appropriate only for enumerating small, but concentrated, organisms like microbes. Towed nets damage fragile jellyfish and comb jellies as they wash to the bottom of the net. Only specialized nets that sample large volumes catch rare animals.

Sampling life in specific environments like coral heads or hydrothermal vents must target small areas with special tools.

The *Multiple Opening/Closing Net and Environmental Sensing System* (MOCNESS) was developed in the 1980s and can be modified for CMarZ to catch small, rare organisms. Its stacked nets open and close at specific depths, sampling a narrow depth range. CMarZ scientists deployed a 10 meter by 10 meter version with fine mesh to capture small organisms. They sampled tens of thousands of cubic meters of water and collected rare deep-sea life, providing the first live observations of some. The mouth of the MOCNESS can be fitted with a high definition video camera and a *Video Plankton Recorder*. The recorder photographs the water entering the net to compare images of live animals entering to damaged ones caught in the net.

Cameras sample the seafloor too. For many decades, bait has attracted deep-sea life in front of cameras, but they quickly eat the bait. Cameras tell us little about how different species respond as day changes to night and seasons change. Release of new bait packages at programmed intervals reveals arrival times and attracts a broader array of life.

Grabs and coring devices evolved from simple beginnings early in the twentieth century to ones that slowly penetrate seafloor sediment without blowing away the tiny animals and microbes at the sediment surface. For decades the devices have collected sediments and their life, and recently the Census studies of continental margins (COMARGE) and abyssal plains (CeDAMar) used them widely.

Because corers cannot penetrate hard bottoms, Census scientists in the Mid-Atlantic Ridge (MAR-ECO) project sampled as fishermen do, trawling rugged bottoms with dredges and fortified nets. Trawling, however, misses many *epifaunal* organisms that live on or attached to the seafloor, and mobile organisms that sense and avoid approaching nets. These shortcomings inspired new submersible tools, towed cameras, ROVs, and AUVs (Figure 3.1).

Early on, Fred Grassle advocated submersibles such as *Alvin* to study the oceans. *Alvin* carries three people into the deep sea and became famous for discovering both hydrothermal vents and the *Titanic*. Grassle recognized the value of first-hand observations that place organisms in the context where they were collected. Leading the first biological expedition to hydrothermal vents in 1979, Grassle needed tools to precisely locate and sample patchy habitats. Vent fields span just a few square kilometers or less of seafloor, and single vent openings are less than a square meter. Because vent organisms require specific habitats within these small areas, sampling even a few centimeters from a vent alters which species are found.

Again and again, scientists have had to adapt to sample the deep sea. The high cost and scarcity of manned submersibles led to the development of unmanned ROVs. Operators watching live video feed relayed through a fiber-optic tether guide the ROV across the seafloor below. Sitting on the ship, they manipulate mechanical hands to collect samples and to deploy and recover experiments. The newest innovation is untethered AUVs programmed to follow a planned path. At the end of a "mission," scientists retrieve the AUV, remove stored data, and recharge the batteries for another excursion.

Submersibles, ROVs, and AUVs collected samples at precise locations for the Census' chemosynthetic ecosystems (ChEss) project. The *Census of Seamounts* (CenSeam) needed similar tools to sample seamount ecosystems quantitatively and without destroying them. ChEss developed AUV tools to track chemical signatures to hydrothermal vent openings and then photograph them on a precise grid. With these tools, scientists have located and mapped black smoker vents and also more diffuse vents, which are trickier to find. Vent temperatures as hot as 407 °C, which melt plastic and damage lenses, make imprecise navigation fatal to expensive equipment.

Specialists working with delicate gelatinous zooplankton recognized that technologies such as scuba diving in shallow water and undersea vehicles in deep water could let human eyes peek at underwater life *in situ*. Census scientists in Japan developed a small ROV system that dives to 1,000 meters depth from small vessels and brings back environmental data and high-quality video of zooplankton.

Ocean imagery

The spectacle of ocean life has inspired artists for millennia. In this tradition, Jacques Perrin and Jacques Cluzaud of Galatée Films partnered with the Census to film *Oceans*. The Census jumped at the opportunity to showcase ocean life to a broad audience. As noted by Jesse Ausubel, "Marine ecology must succeed both as an applied science and like astronomy as a source of wonder and humility. I firmly support an alliance between art and science." For four years, Census scientists from around the world collaborated with the Galatée team while they filmed 200 species at more than 50 locations globally.

The Galatée team captured the beauty, speed, fragility, and strength of ocean life with new technologies. In rough seas, stabilized cameras removed vibration, and an electric mini-helicopter approached and filmed animals quietly. A towed torpedo camera moved with animals swimming as fast as 15 knots, and a camera on a pole over the side of vessels filmed others swimming at 8 knots. Collectively, these techniques showed animals moving in ways not previously seen. For the public, the result was a visually spectacular film. For science, the Galatée team created a scientific legacy of more than 400 hours of images of organisms moving in their natural habitats.

Listening to the ocean

Acoustic sensors send out sound waves that bounce off objects and show their outlines. Acoustics now map topography and sediment type that defines seafloor habitat and species. Some acoustic tools can map seafloor biota such as coral or clam beds, and thus guide sampling. After acoustic surveys, sampling can target particular regions and extrapolate the results more broadly for management applications.

Acoustics measure abundance of ocean life, one of three main objectives of the Census. Gas-filled swim bladders in fish control their buoyancy and thus their depth without wasting valuable energy swimming. Swim bladders also reflect sound waves in a distinctive way so individual fish can be counted without lowering a net into the water. The reflected sound helps managers decide how many fish remain and how many can be caught before a fishery becomes unsustainable. Formerly the managers depended heavily on historical trends of catch by fishermen, which is more extensive, but less reliable, than scientific trawl and acoustic surveys. Improving fishing technology can counter declining trends in catch, so populations may appear stable or even growing when the opposite is true. Scientific trawling surveys can correct for changing fishing effort, but they are expensive and slow.

Until recently, acoustic surveys also proceeded slowly and estimated abundances in small spans of ocean. But acoustic echosounders now widely and swiftly survey a 100-square-kilometer swath of seabed bigger than Manhattan in less than a minute. Scientists used this technology off

Figure 4.2 Tagged animals make oceanographic observations from an animal's perspective.
The range of tags now available potentially allows scientists to eventually track three-quarters of the fish and mammal species as they swim around Planet Ocean. Smart tags also measure depth, temperature, and light. As they swim and dive, elephant seals, for example, collect temperature observations in the Southern Ocean. The tag on the head of the female southern elephant seal pictured on Livingston Island, Antarctica, stores observations on her environment for scientists to read later.

New Jersey and then on Georges Bank and counted approximately 250 million fish (50,000 tons) in one large school, probably herring, without taking any from the ocean or harming them (Figure 4.1). Although different species may reflect sound in similar, indistinguishable ways, when one species dominates, acoustics are groundbreaking for censusing marine life with no fixed address.

Acoustics also provided submarine communication for the Census. Hydrophones count salmon migrating past underwater listening stations along the Pacific coast of North America and acoustic tags follow diving mammals in the Antarctic and elsewhere (Figure 4.2). Acoustics allow us to see the oceans as animals do.

Where animals move, rest, feed, and reproduce

The variability of Planet Ocean determines who lives where, and when and why they move. Distribution of marine life, along with diversity and abundance, inspired the Census. Fish and mammals move hundreds or even thousands of kilometers in pursuit of mates, food, suitable temperatures, or oxygen. All burn energy and risk death by predation and starvation.

Species that breathe air at the surface often dive to dine. Seals and whales dive from the surface where they breathe down hundreds or even thousands of meters to find prey with whistling and sonar. Fish, crustaceans, and other invertebrates hide in the dark deep where predators can't see them, but then surface at night to feed. For fishes and invertebrates, these vertical migrations create a dawn and dusk "rush hour," where predators and prey commute to and from their preferred deep locations to rest or dine safely near the surface at night. Some move up from 500–1,000 meter daytime depths, rising higher than elevators carry humans in the tallest skyscrapers. The octopus *Stauroteuthis syrtensis* lives near the bottom, but has been collected 1,690 meters above the seafloor.

Horizontal migrations often follow familiar, short routes. Blue crabs travel only a few kilometers, moving from the coastal ocean into estuaries to spawn. At the other extreme, whales, turtles, and seabirds swim or fly thousands of kilometers from areas to feed and grow to other locations where they mate and give birth. The juvenile and adolescent turtles that hatch in the western Pacific swim to California to feed, almost matching the long migrations of bluefin tuna. Turtles can't match the swimming speeds of Olympic champions like Michael Phelps, but they swim farther. Still, bluefin tuna racing faster than 90 kilometers per hour wallop the 7 kilometers per hour that won Phelps a gold medal in Beijing.

The movements of many species have been known in general terms for more than a century. Captains on whaling ships knew where to locate their targets at different times of year from ship observations around the world. For decades, scientists attached small tags to fish, offering rewards for returning and reporting their location. Some tags were returned, but what about those that are never seen again and where had the returned tags traveled before they were found?

Like other technology, tagging has recently progressed by leaps and bounds. *Passive Integrated Transponder* (PIT) tags emit radio frequencies akin to those attached to merchandise in stores to foil shoplifters or track inventory.

When signaled, a tag responds without harming the fish. Unfortunately, the fish must pass within tens of centimeters of the receiving scanner. Nonetheless, for decades PIT tag technology has counted salmon swimming across river dams, an ideal application because they funnel through a narrow fish ladder.

Since the late 1950s, scientists have implanted acoustic tags in fish that signal receivers as fish swim past. Their acoustic signals extend hundreds of meters, a limited range, but farther than the radio signals of PIT tags. The pencil eraser size and 2–3 year battery are further improvements.

The *Pacific Ocean Shelf Tracking Project* (POST) of the Census arrayed "listening curtains" of hydrophones at strategic places to identify tagged fishes as they passed. The arrays were designed to survive storms and trawlers. The proportion of tagged salmon and sturgeon that returned to spawn, swimming past curtains at pinch points at the mouths of rivers or between Vancouver Island and the mainland, revealed survival rates. Fish passing each array showed the movement of multiple species of salmon, green sturgeon, and six-gill shark. Knowing their movement helps us know their growth and age at reproduction. Although these approaches requiring listening curtains that help little in tracking migration in the open ocean where no land masses create bottlenecks, they are powerful tools for identifying and tracking *anadromous* fish that migrate from the ocean into rivers to spawn.

Animal ocean view

The "monkeycam" mounted on a monkey's back reveals an amusing "monkey view" of the world on David Letterman's television show. Since the 1960s, marine biologists mounted similar cameras on animals for serious "*biologging*" of the environment wherever the animal swims or dives. The earliest depth recorders attached to seals timed impressive diving feats. For example, northern elephant seals routinely dive down to 600 meters and occasionally to 1,550 meters. Since then, miniaturized, more reliable biologgers have increased the range of sensors and the migrations that they track. Satellites now communicate with the animal's sensors from great distances.

The *Tagging of Pacific Predators* (TOPP) project refined tags to measure temperature, light, depth, and salinity around animals and refined other tags to sense their pulse and temperature. Biologging now brings together behavior, physiology, and oceanography to show how an animal experiences its environment. Sensors that measure temperature, salinity, and depth are attached to seabirds, fishes, and marine mammals (Figure 4.2). Light and pressure sensors on tags estimate chlorophyll, and thus phytoplankton abundance. From light and depth, new algorithms calculate longitude, while latitude is estimated from the timing of sunrise and sunset.

Specialized tags have specialized uses. Archival tags that inertly accompany animals and write an electronic diary can track species like elephant seals or shearwaters that return to colonies where their tags can be recovered. For real-time tracking, TOPP developed GPS tags that relay observations and exact position to satellites. Because diving animals at depth cannot communicate with satellites while they are in deep water, tags on whales and seals broadcast as animals surface to breathe. When Penelope, the elephant seal, swims 14,400 kilometers over a 7-month period, for example, we know exactly where she is and much about her experiences. And even late-night television viewers have been following the travels of "Stelephant Colbert," an elephant seal named for American satirical news host Stephen Colbert.

For animals that breathe oxygen in water, and do not surface to breathe as marine mammals must, new "popup" tags store data and then detach from the animal and float to the surface to relay data through satellites back to land. These tags, which don't have to be retrieved, collect and then transmit biological and environmental data from sharks and tuna.

Biologging innovations come fast, promising to see the ocean as animals do. For many years, the *conductivity–temperature–depth* (CTD) sensors lowered over the sides of ships have precisely measured salinity, temperature, and depth. Now attached to animals, biologgers explore ocean fronts and sea ice formation for us. TOPP developed biologging technology to map the physical ocean to find migration "highways" and biological hot spots where animals congregate. While improving biologging knowledge, it also guides management. Interviewed on *The Today Show*, a US morning news program, TOPP co-leader Barbara Block explained, "This is a wild ocean . . . we have an opportunity in our lifetime to work in this ocean to protect it."

Because even miniaturized archival tags are still too bulky to attach to small animals such as young salmon, they mostly track marine mammals, large fishes, and birds. But the first computers filled large rooms and did far less than the Blackberries and iPods that fit in our hands today. Some day, small tags will move with small fish and even smaller ocean life.

Today, tags reckon location from day length and the timing of sunset, but too imprecisely for complex coasts where precise location is critical. Researchers working in northern Europe inserted archival tags in fish, but learned that in seas with midnight summer sun, the tags could not calculate location! Until new technologies and miniaturization resolve these limits to knowledge, radio tags will remain the technology of choice in small animals and in small, complex coastal regions.

Making the most of what we already measure

Mistaken identity confuses the analysis of measurements already on the books. Species that look alike despite vital biological differences demand tools to distinguish them. The Census has embraced two emerging molecular technologies, barcoding and pyrosequencing, that push back this limit to knowledge. The *Marine Barcode of Life* (MarBOL) has collaborated with Census projects to identify species using barcodes that differ a lot between species, but little if at all within a species.

Project CMarZ built a seagoing assembly line to barcode many of the 7,000 currently known zooplankton species. Taxonomists removed individual specimens from the net, identified them, and then barcoded them in a shipboard molecular lab. Much like an automobile assembly line, this efficient marine assembly line ensures samples are fresh and links a unique barcode to an established name based on morphology. CMarZ, like ICoMM, the *International Census of Marine Microbes*, has also begun bulk genetic identification without sorting individuals. Mixing biodiversity in a blender speeds processing and skips tedious microscope work. Although the blend cannot accurately tell us abundances of species, the presence or absence of known barcodes does show presence or absence of species. Already DNA sensors detect toxic algal blooms and specific larvae of invertebrate bivalves. Although COI barcoding cannot discriminate among microbes unless they have mitochondria, sequencing that determines the order of the nucleotides that compose DNA and RNA found in the other parts of the cell forms the core tool developed by ICoMM. Pyrosequencing uses chemoluminescence of the nucleotide sequences of other targets such as short loops of ribosomal RNA particularly well. Pyrosequencing identifies microbes from bacteria to archaea to protists. The integration of reference DNA barcodes for multicellular organisms and some microbes and RNA loops in many microbes can populate genetic "libraries" that can be paired with classic taxonomy to make the most of what we have already measured.

Census projects have made consistent methodologies a cornerstone of their large and intercomparable datasets to build on previous broad-scale comparisons of drifting plankton or seafloor benthos. The *Census of Coral Reef Ecosystems* project (CReefs) developed tools to compare reefs. The Autonomous Reef Monitoring Structure (ARMS) is a stack of standard plates that mimic reefs. They deploy the "reef condos" around the world and, after some time, recover them to find who and how many have moved in, providing a standardized comparison of colonizers. Because some organisms avoid ARMS, they sample incompletely, but they show relative diversity without methodological complications.

Other Census projects have compared existing standard methods at multiple locations. The *Natural Geography in Shore Areas* (NaGISA) project sampled transects in exactly the same way to produce a standardized global dataset on life from shores out to about 20 meters depth. NaGISA wrote a methodology guide for intertidal studies, and the *Arctic Ocean Diversity* (ArcOD) project also wrote one for ice researchers. For deep-sea sampling, the *Census of Diversity of Abyssal Marine Life* (CeDAMar) wrote on methods to ensure comparability of samples.

Census scientists contributed to standard checklists of species for seamount fishes, Chinese ocean biota, Gulf of Mexico biodiversity, and Japanese zooplankton, and field guides for Californian and global hydrozoans. They also produced regional guides to fauna and flora, as well as keys to identify squat lobster. CMarZ scientist Vijayalakshmi Nair, a retired scientist in India, brought retired taxonomists back to their microscopes, linking their expertise to barcodes, ensuring that their classic knowledge of species would link to barcode data. Together, these strategies shared data, even between generations of experts, and thus standardized protocols for valid comparisons.

Sharing fosters collaboration

The Census deposits all its data in the *Ocean Biogeographic Information System* (OBIS) to share with everyone via the World Wide Web. Census projects have developed other databases and analytical tools, many linked to OBIS. OBIS verifies its entries with its partner, the *World Register of Marine*

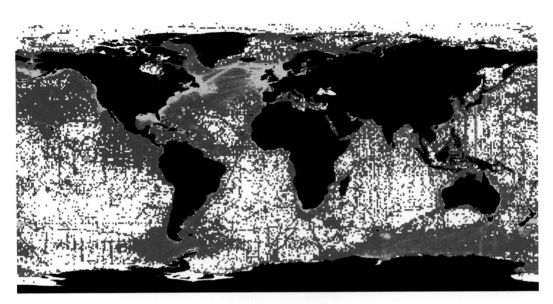

Number of Records

Low   High

**Figure 4.3 The Ocean Biogeographic Information
System (OBIS) makes information accessible.**
OBIS unites the Census, providing a universally accessible database about ocean
life. As of 2009, OBIS had grown to 22 million records from more than 700 different
databases integrated into one. The map shows the global distribution of OBIS data
records, with a color scale ranging from blue dots for few to red for many
records, thus highlighting locations of sampling strengths and gaps.

Species (WoRMS), which is compiling an authoritative global list of all
marine species names. By October 2010, viewers will find nearly 30 million
data records at www.iobis.org (Figure 4.3). Each record identifies geographic
position, depth, collection date, source, and verified species name of each
specimen. Analysts can overlay global distributions of species and diversity
mapped on environmental drivers like water temperature or salinity.

   The 17 projects within the Census pull together massive datasets,
some brand new and some assembled from sources scattered around the
world. Merging the datasets and then displaying them with new visualization
tools in novel ways evokes fresh, intuitive understanding. The *Mapping and
Visualization* team of the Census coordinated with the *Education and*

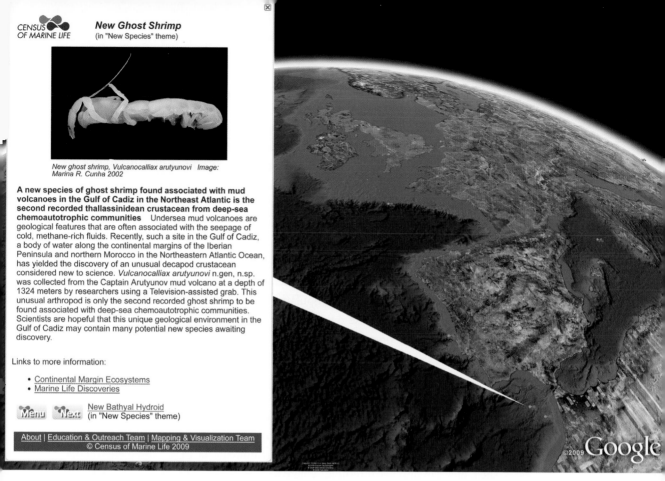

Figure 4.4 Visualization makes the information public.
The Census is part of Google Earth. There, any viewer of the Ocean layer can find Census pages for some of its newly discovered species, such as the ghost shrimp shown here, and other information on ocean life.

Outreach group and Census projects to develop a Census highlight layer, now included within the Google Earth application (Figure 4.4). New databases spread before audiences a new view on maritime boundaries, human impacts, and biogeographical provinces. Collaborating with the *National Geographic Society*, the Mapping and Visualization team is drawing and distributing a wall map that will include Census activities and discoveries (see Figure 1.4).

Analysis of the frequent positions of TOPP's tagged animals demonstrates how to make the most of what we already have despite imperfect observations. Models of the FMAP project analyze observations over time to locate an animal, despite uncertainties about positions. The models weight each position to filter out unlikely locations and weight closer locations more heavily than unreasonable ones. For example, gray seals don't leap unreasonably around the ocean, thus their migration can be inferred more precisely by filtering out locations that imply great leaps. A second model infers complex behaviors along migration routes. Leatherback turtles migrating in the North Atlantic slow down at night to feed or navigate cautiously. Differentiating between an animal migrating and stopping to dine tells analysts where animals are and why during their long migrations. These models pinpoint subtle environmental cues that replace signposts in the open ocean.

Pivotal thresholds or *reference points* in population numbers tell biologists how many individuals of a species can be fished without spiraling the population toward collapse. An underexploited fishery sheds jobs and leaves nations hungry. Overexploited populations recover slowly, if at all. But better statistical estimates of these reference points are now possible, including how many fish are present and how much fishing a population can endure.

For over a century ecologists have compared ecosystems with *species-area curves* to show how quickly new species are found with wider sampling, and more recently to evaluate species loss. FMAP proposed that finding species more slowly as sampling of coral reefs widened showed fishing impact. Because unseen species can mislead, this approach must be used with caution.

Statistical tools predict broad pattern from limited data. In measurements on the shelf, they can find key thresholds where temperature, depth, or bottom substrate limits the distributions of species. Thresholds, often in easily measured variables, tell where species are likely to be found. For example, depth, temperature, and carbonate predict where deep-water corals live on seamounts.

Great auks, Steller sea cows, and Caribbean monk seals were visibly hunted to extinction, but statistical tools may help us know when less visible species have been lost. One study suggested that of 133 candidate marine species, only 21 were globally extinct rather than regionally eliminated. Extinctions included three marine mammals, five marine birds, three marine fishes, five marine mollusks, three corals/anemones, and two types of algal species. Although many consider broadly distributed ocean species less vulnerable than their terrestrial and freshwater counterparts, this assumption is sometimes incorrect because marine extinctions are harder to see.

Once a species is no longer seen we may think it has gone extinct, but wider searching and new methods may rediscover it later. Fishermen off South Africa caught the famous coelacanth, once thought extinct. How far a search must be extended to find a species can tell us if its habitat is shrinking or lost. Species that occur in a specific habitat and are sampled irregularly make

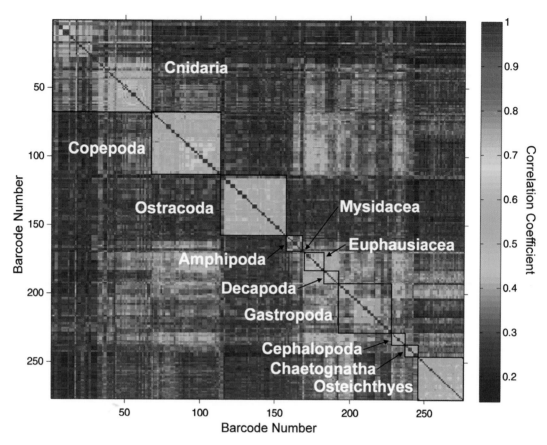

Figure 4.5 Genetics reveals the evolutionary connections of life in Planet Ocean.
Genetic barcode groupings of life show sharp edges for some groups and closer
relationships for others. The image shows vector analysis "Klee diagrams" used by the
Census. Red and yellow show species are more closely related to each other than
to distant blue species. See Chapter 7.

this approach difficult. Incomplete sampling of a site in different seasons or
sampling during natural environmental fluctuations can exaggerate depletion
or obscure extinction. In short, good biological data are needed to infer
extinction, and shortcuts are few.

Statistical tools can show novel evolutionary relationships. With vector
analysis of many barcode data, CMarZ diagrammed similarity among
zooplankton taxa. Because the diagrams resemble geometrical paintings by
the twentieth-century Swiss painter Paul Klee, they are called "Klee diagrams."
The diagrams can be viewed at a range of resolutions "zooming in" on a
specific region like a single genus to compare species relationships (Figure 4.5).

ICoMM inferred evolutionary relationships in living microbes from the clues of fossilized lipids that persist in sediments. This emerging approach promises new knowledge of how microbes are related, a difficult problem in a group with few morphological features.

The Census has invented, adapted, and embraced these tools to extract a new view of the living ocean and its diversity, distribution, and abundance. Beginning around the ocean rim, exploration of Planet Ocean continues.

BIBLIOGRAPHY

Adams, P. B., C. Grimes, J. E. Hightower, *et al.*, Population status of North American green sturgeon, *Acipenser medirostris. Environ. Biol. Fishes*, **79**:3–4 (2007), 339–56.

Amaral-Zettler, L., L. F. Artigas, J. Baross, *et al.*, A global census of marine microbes. In A. D. McIntyre (ed.), *Life in the World's Oceans: Diversity, Distribution, and Abundance* (Oxford: Blackwell Publishing Ltd., 2010), pp. 223–45.

Amaral-Zettler, L. A., E. A. McCliment, H. W. Ducklow, and S. M. Huse, A method for studying protistan diversity using massively parallel sequencing of V9 hypervariable regions of small-subunit ribosomal RNA genes. *PLoS ONE*, **4**:7 (2009).

Anderson, D. M., D. M. Kulis, B. A. Keafer, *et al.*, Identification and enumeration of *Alexandrium spp.* from the Gulf of Maine using molecular probes. *Deep Sea Res. II*, **52**:19–21 (2005), 2467–90.

Andrews, K. S., P. S. Levin, S. L. Katz, *et al.*, Acoustic monitoring of sixgill shark movements in Puget Sound: evidence for localized movement. *Can. J. Zool.*, **85**:11 (2007), 1136–42.

Andrews, K. S., G. D. Williams, D. Farrer, *et al.*, Diel activity patterns of sixgill sharks, *Hexanchus griseus*: the ups and downs of an apex predator. *Anim. Behav.*, **78**:2 (2009), 525–36.

Appeltans, W., P. Bouchet, G. A. Boxshall, *et al.*, World Register of Marine Species (2010). Available from: www.marinespecies.org.

Auster, P. J., J. Lindholm, M. Plourde, K. Barber, and H. Singh, Camera configuration and use of AUVs to census mobile fauna. *Mar. Technol. Soc. J.*, **41**:4 (2007), 49–52.

Baba, K., E. Macpherson, G. C. B. Poore, *et al.*, Catalogue of squat lobsters of the world (Crustacea: Decapoda: Anomura – families Chirostylidae, Galatheidae and Kiwaidae). *Zootaxa*, **1905** (2008), 3–220.

Bailey, D. M., M. A. Collins, J. D. M. Gordon, A. F. Zuur, and I. G. Priede, Long-term changes in deep-water fish populations in the northeast Atlantic: a deeper reaching effect of fisheries? *Proc. R. Soc. Lond. B Biol. Sci.*, **276**:1664 (2009), 1965–9.

Baker, M. C., E. Ramirez Llodra, P. A. Tyler, *et al.*, Biogeography, ecology and vulnerability of chemosynthetic ecosystems in the deep sea. In A. D. McIntyre (ed.), *Life in the World's Oceans: Diversity, Distribution, and Abundance* (Oxford: Blackwell Publishing Ltd., 2010), pp. 161–82.

Bekker-Nielsen, T. (ed.), *Ancient Fishing and Fish Processing in the Black Sea Region* (Oxford: Aarhus University Press, 2006).

Beaugrand, G., P. C. Reid, F. Ibanez, and B. Planque, Biodiversity of North Atlantic and North Sea calanoid copepods. *Mar. Ecol. Prog. Ser.*, **204** (2000), 299–303.

Block, B. A., Physiological ecology in the 21st century: advancements in biologging science. *Integr. Comp. Biol.*, **45**:2 (2005), 305–20.

Block, B. A., D. P. Costa, and S. J. Bograd, A view of the ocean from Pacific predators. In A. D. McIntyre (ed.), *Life in the World's Oceans: Diversity, Distribution, and Abundance* (Oxford: Blackwell Publishing Ltd., 2010), pp. 291–311.

Block, B. A., H. Dewar, S. B. Blackwell, *et al.*, Migratory movements, depth preferences, and thermal biology of Atlantic bluefin tuna. *Science*, **293**:5533 (2001), 1310–4.

Block, B. A., H. Dewar, C. J. Farwell, and E. D. Prince, A new satellite technology for tracking the movements of Atlantic bluefin tuna. *Proc. Natl. Acad. Sci. USA*, **95** (1998), 9384–9.

Boustany, A. M., R. Matteson, M. R. Castleton, C. J. Farwell, and B. A. Block, Movements of Pacific bluefin tuna (*Thunnus orientalis*) in the Eastern North Pacific revealed with archival tags. *Prog. Oceanogr.*, **86** (2010), 94–104.

Bradford-Grieve, J. M., *Hyperbionyx athesphatos* n.sp. (Calanoida: Hyperbionychidae), a rare deep-sea benthopelagic species taken from the tropical North Atlantic. *Deep Sea Res. II* (in press).

Breed, G. A., W. D. Bowen, J. I. McMillan, and M. L. Leonard, Sexual segregation of seasonal foraging habitats in a non-migratory marine mammal. *Proc. R. Soc. Lond. B Biol. Sci.*, **273**:1599 (2006), 2319–26.

Bucklin, A. C., S. Nishida, S. Schnack-Schiel, *et al.*, A census of zooplankton of the global ocean. In A. D. McIntyre (ed.), *Life in the World's Oceans: Diversity, Distribution, and Abundance* (Oxford: Blackwell Publishing Ltd., 2010), p. 247.

Charrassin, J. B., M. Hindell, S. R. Rintoul, *et al.*, Southern Ocean frontal structure and sea-ice formation rates revealed by elephant seals. *Proc. Natl. Acad. Sci. USA*, **105**:33 (2008), 11634–9.

Collins, T., F. Gorell, and D. T. Crist, Diverse sea "bugs", other life revealed on landmark Atlantic cruise to census zooplankton and animals at depths to 5km. 2006. Available from: www.coml.org/comlfiles/press/Final-Census-news-release-05–04.pdf.

CoML, Oceans. 2010. Available from: http://www.coml.org/comlfiles/press/Oceans_Film_Brochure_Public1.pdf.

Consalvey, M., M. R. Clark, A. A. Rowden, and K. I. Stocks, Life on seamounts. In A. D. McIntyre (ed.), *Life in the World's Oceans: Diversity, Distribution, and Abundance* (Oxford: Blackwell Publishing Ltd., 2010), pp. 123–38.

Costa, D. P. and B. Sinervo, Field physiology: physiological insights from animals in nature. *Annu. Rev. Physiol.*, **66** (2004), 209–38.

Crocker, D. E., D. P. Costa, B. J. Le Boeuf, P. M. Webb, and D. S. Houser, Impact of El Niño on the foraging behavior of female northern elephant seals. *Mar. Ecol. Prog. Ser.*, **309** (2006), 1–10.

Decker, C. J. and C. Reed, The National Oceanographic Partnership Program, a decade of impacts on oceanography. *Oceanography*, **22**:2 (2009), 208–27.

Diaz, J. M., F. Gast, and D. C. Torres, Rediscovery of a Caribbean living fossil: *Pholadomya candida* GB Sowerby I, 1823 (Bivalvia: Anomalodesmata: Pholadomyoidea). *Nautilus*, **123**:1 (2009), 19–20.

Dulvy, N. K., Y. Sadovy, and J. D. Reynolds, Extinction vulnerability in marine populations. *Fish Fish.*, **4**:1 (2003), 25–64.

Dupre, S., G. Buffet, J. Mascle, *et al.*, High-resolution mapping of large gas emitting mud volcanoes on the Egyptian continental margin (Nile Deep Sea Fan) by AUV surveys. *Mar. Geophys. Res.*, **29**:4 (2008), 275–90.

Ebbe, D. S., M. Billett, A. Brandt, *et al.*, Diversity of abyssal marine life. In A. D. McIntyre (ed.), *Life in the World's Oceans: Diversity, Distribution, and Abundance* (Oxford: Blackwell Publishing Ltd., 2010), pp. 139–60.

Eicken, H., A. L. Lovecraft, and M. L. Druckenmiller, Sea-ice system services: a framework to help identify and meet information needs relevant for arctic observing networks. *Arctic*, **62**:2 (2009), 119–36.

Ekstrom, P. A., An advance in geolocation by light. *Mem. Natl. Inst. Polar Res. (Tokyo)*, **58** (2004), 210–26.

Erickson, D. L. and M. A. H. Webb, Spawning periodicity, spawning migration, and size at maturity of green sturgeon, *Acipenser medirostris*, in the Rogue River, Oregon. *Environ. Biol. Fishes*, **79**:3–4 (2007), 255–68.

Fedak, M., P. Lovell, B. McConnell, and C. Hunter, Overcoming the constraints of long range radio telemetry from animals: getting more useful data from smaller packages. *Integr. Comp. Biol.*, **42**:1 (2002), 3–10.

Felder, D. L. and D. K. Camp (eds.), *Gulf of Mexico: Origins, Waters, and Biota: Volume 1, Biodiversity* (Texas: A&M University Press, 2009).

Flemming, J. E. M., C. A. Field, M. C. James, I. D. Jonsen, and R. A. Myers, How well can animals navigate? Estimating the circle of confusion from tracking data. *Environmetrics*, **17**:4 (2006), 351–62.

Foote, K. G., Pilot census of marine life in the Gulf of Maine: contributions of technology. *Oceanol. Acta*, **25**:5 (2002), 213–8.

Frederick, J. L., Post-settlement movement of coral reef fishes and bias in survival estimates. *Mar. Ecol. Prog. Ser.*, **150**:1–3 (1997), 65–74.

Froese, R. and A. Sampang, Taxonomy and biology of seamount fishes. In T. Morato and D. Pauly (eds.), *Seamounts: Biodiversity and Fisheries* (Fisheries Centre Research Reports, 12:5, 2004), pp. 25–32.

German, C. R., E. T. Baker, D. P. Connelly, *et al.*, Hydrothermal exploration of the Fonualei Rift and Spreading Center and the Northeast Lau Spreading Center. *Geochem. Geophys. Geosyst.*, **7** (2006), doi:10.1029/2006GC001324.

German, C. R., S. A. Bennett, D. P. Connelly, *et al.*, Hydrothermal activity on the southern Mid-Atlantic Ridge: tectonically- and volcanically-controlled venting at 4–5° S. *Earth Planet Sci. Lett.*, **273**:3–4 (2008), 332–44.

German, C. R., D. R. Yoerger, M. Jakuba, *et al.*, Hydrothermal exploration with the Autonomous Benthic Explorer. *Deep Sea Res. I*, **55**:2 (2008), 203–19.

Gibson, A. J. F. and R. A. Myers, Estimating reference fishing mortality rates from noisy spawner-recruit data. *Can. J. Fish. Aquat. Sci.*, **61**:9 (2004), 1771–83.

Gilly, W. F., U. Markaida, C. H. Baxter, *et al.*, Vertical and horizontal migrations by the jumbo squid *Dosidicus gigas* revealed by electronic tagging. *Mar. Ecol. Prog. Ser.*, **324** (2006), 1–17.

Grassle, J. F., The ecology of deep-sea hydrothermal vent communities. *Adv. Mar. Biol.* (1986), 301–62.

Grassle, J. F., H. L. Sanders, R. R. Hessler, G. T. Rowe, and T. McLellan, Pattern and zonation: a study of the bathyal megafauna using the research submersible Alvin. *Deep-Sea Res. Oceanographic Abstr.*, **22**:7 (1975).

Greene, C. H., B. A. Block, D. Welch, *et al.*, Advances in conservation oceanography new tagging and tracking technologies and their potential for transforming the science underlying fisheries management. *Oceanography*, **22**:1 (2009), 210–23.

Gunn, J., A. Rogers, and E. Urban, Observation of ocean biology on a global scale: is new technology required for Bio-GOOS? In J. Hall, D. E. Harrison, and D. Stammer (eds.), *Proceedings of OceanObs'09: Sustained Ocean Observations and Information for Society*, Vol. **1**, Venice, Italy, 21–25 September 2009 (ESA Publication WPP-306, 2009).

Haase, K. M., A. Koschinsky, S. Petersen, *et al.*, Diking, young volcanism and diffuse hydrothermal activity on the southern Mid-Atlantic Ridge: the Lilliput field at 9°33'S. *Mar. Geol.*, **266**:1–4 (2009), 52–64.

Hamner, W. M., Underwater observations of blue-water plankton – logistics, techniques, and safety procedures for divers at sea. *Limnol. Oceanogr.*, **20**:6 (1975), 1045–50.

Harley, S. J., R. A. Myers, and C. A. Field, Hierarchical models improve abundance estimates: spawning biomass of hoki in Cook Strait, New Zealand. *Ecol. Appl.*, **14**:5 (2004), 1479–94.

Hebert, P. D. N., A. Cywinska, S. L. Ball, and J. R. DeWaard, Biological identifications through DNA barcodes. *Proc. R. Soc. Lond. B Biol. Sci.*, **270**:1512 (2003), 313–21.

Heirtzler, J. R. and J. F. Grassle, Deep sea research by manned submersibles. *Science*, **194**:4262 (1976), 294–9.

Hill, R. D. and M. J. Braun, Geolocation by light level – the next step: latitude. In J. R. Sibert and J. G. Nielsen (eds.), *Electronic Tagging and Tracking in Marine Fisheries* (The Netherlands: Kluwer Academic Publishers, 2001), pp. 315–30.

Huber, J. A., D. B. Mark Welch, H. G. Morrison, *et al.*, Microbial population structures in the deep marine biosphere. *Science*, **318**:5847 (2007), 97–100.

Hutchings, J. A., Collapse and recovery of marine fishes. *Nature*, **406**:6798 (2000), 882–5.

Isaacs, J. D., The nature of oceanic life. *Sci. Am.*, **221**:3 (1969), 146–62.

Jackson, J. B. C., Ecological extinction and evolution in the brave new ocean. *Proc. Natl. Acad. Sci. USA*, **105**:Suppl. 1 (2008), 11458–65.

Johnson, G. D., J. R. Paxton, T. T. Sutton, *et al.*, Deep-sea mystery solved: astonishing larval transformations and extreme sexual dimorphism unite three fish families. *Biol. Lett.*, **5**:2 (2009), 235–9.

Johnson, J. H., Sonic tracking of adult salmon at Bonneville Dam, 1957. *Fish. Bull.*, **176** (1960).

Jonsen, I. D., J. M. Flemming, and R. A. Myers, Robust state-space modeling of animal movement data. *Ecology*, **86**:11 (2005), 2874–80.

Jonsen, I. D., R. A. Myers, and M. C. James, Robust hierarchical state-space models reveal diel variation in travel rates of migrating leatherback turtles. *J. Anim. Ecol.*, **75**:5 (2006), 1046–57.

Jonsen, I. D., R. A. Myers, and M. C. James, Identifying leatherback turtle foraging behavior from satellite telemetry using a switching state-space model. *Mar. Ecol. Prog. Ser.*, **337** (2007), 255–64.

Jorgensen, S. J., J. Reeb, T. Chapple, *et al.*, Philopatry and migration of Pacific white sharks. *Proc. R. Soc. Lond. B Biol. Sci.*, **277** (2010), 679–88.

Kemp, K. M., A. J. Jamieson, P. M. Bagley, M. A. Collins, and I. G. Priede, A new technique for periodic bait release at a deep-sea camera platform: first results from the Charlie-Gibbs Fracture Zone, Mid-Atlantic Ridge. *Deep Sea Res. II*, **55**:1–2 (2008), 218–28.

Kitamura, M., D. Lindsay, H. Miyake, and T. Horita, Ctenophora. In K. Fujikura, T. Okutani, and T. Maruyama (eds.), *Deep-sea Life – Biological Observations using Research Submersibles* (Kanagawa: Tokai University Press, 2008).

Kitamura, M., H. Miyake, and D. Lindsay, Cnidaria. In K. Fujikura, T. Okutani, and T. Maruyama (eds.), *Deep-sea Life – Biological Observations using Research Submersibles* (Kanagawa: Tokai University Press, 2008), pp. 295–320.

Knowlton, N., R. E. Brainard, R. Fisher, *et al.*, Coral reef biodiversity. In A. D. McIntyre (ed.), *Life in the World's Oceans: Diversity, Distribution, and Abundance* (Oxford: Blackwell Publishing Ltd., 2010), pp. 65–77.

Konar, B., K. Iken, G. Pohle, *et al.*, Surveying nearshore biodiversity. In A. D. McIntyre (ed.), *Life in the World's Oceans: Diversity, Distribution, and Abundance* (Oxford: Blackwell Publishing Ltd., 2010), pp. 27–41.

Kooyman, G. L., Genesis and evolution of bio-logging devices: 1963–2002. *Mem. Natl. Inst. Polar Res. Spec. Issue*, **58** (2004), 15–22.

Koschinsky, A., Discovery of new hydrothermal vents on the southern Mid-Atlantic Ridge (4°–10°) during cruise M68/1. *InterRidge News*, **15** (2006), 9–15.

Koschinsky, A., D. Garbe-Schanberg, S. Sander, *et al.*, Hydrothermal venting at pressure-temperature conditions above the critical point of seawater, 5°S on the Mid-Atlantic Ridge. *Geology*, **36**:8 (2008), 615–18.

Lawson, G. L. and G. A. Rose, Small-scale spatial and temporal patterns in spawning of Atlantic cod (*Gadus morhua*) in coastal Newfoundland waters. *Can. J. Fish. Aquat. Sci.*, **57**:5 (2000), 1011–24.

Lindley, S. T., M. L. Moser, D. L. Erickson, *et al.*, Marine migration of North American green sturgeon. *Trans. Am. Fish. Soc.*, **137**:1 (2008), 182–94.

Lindsay, D., A checklist of midwater cnidarians and ctenophores from Sagami Bay – species sampled during submersible surveys from 1993–2004. *Bull. Plankton Soc. Japan*, **53** (2006), 104–10.

Lindsay, D. J. and J. C. Hunt, Biodiversity in midwater cnidarians and ctenophores: submersible-based results from deep-water bays in the Japan Sea and north-western Pacific. *J. Mar. Biol. Assoc. UK*, **85**:3 (2005), 503–17.

Lindsay, D. J. and H. Miyake, A novel benthopelagic ctenophore from 7,217 m depth in the Ryukyu Trench, Japan, with notes on the taxonomy of deep-sea cydippids. *Plankton and Benthos Res.*, **2**:2 (2007), 98–102.

Lindsay, D. and H. Miyake, A checklist of midwater cnidarians and ctenophores from Japanese waters – species sampled during submersible surveys from 1993–2008 with notes on their taxonomy. *Kaiyo Monthly*, **41** (2009), 417–38.

Liu, J. Y., *Checklist of Marine Biota of China Seas* (Institute of Oceanology, Chinese Academy of Sciences, 2008).

Lotze, H. K. and M. Glaser, Ecosystem services of semi-enclosed marine systems. In E. R. Urban, B. Sundby, P. Malanotte-Rizzoli, and J. M. Melillo (eds.), *Watersheds, Bays and Bounded Seas* (Washington, DC: Island Press, 2008), pp. 227–49.

Lutcavage, M. E., R. W. Brill, G. B. Skomal, B. C. Chase, and P. W. Howey, Results of pop-up satellite tagging of spawning size class fish in the Gulf of Maine: do North Atlantic bluefin tuna spawn in the mid-Atlantic? *Can. J. Fish. Aquat. Sci.*, **56**:2 (1999), 173–7.

Machida, R. J., Y. Hashiguchi, M. Nishida, and S. Nishida, Zooplankton diversity analysis through single-gene sequencing of a community sample. *BMC Genomics*, **10** (2009).

Makris, N. C., P. Ratilal, S. Jagannathan, *et al.*, Critical population density triggers rapid formation of vast oceanic fish shoals. *Science*, **323**:5922 (2009), 1734–7.

Makris, N. C., P. Ratilal, D. T. Symonds, *et al.*, Fish population and behavior revealed by instantaneous continental shelf-scale imaging. *Science*, **311**:5761 (2006), 660–3.

Medici, D. A., T. G. Wolcott, and D. L. Wolcott, Scale-dependent movements and protection of female blue crabs (*Callinectes sapidus*). *Can. J. Fish. Aquat. Sci.*, **63** (2006), 858–71.

Melchert, B., C. W. Devey, C. R. German, *et al.*, First evidence for high-temperature off-axis venting of deep crustal/mantle heat: the Nibelungen hydrothermal field, southern Mid-Atlantic Ridge. *Earth Planet Sci. Lett.*, **275**:1–2 (2008), 61–9.

Melnychuk, M. C., Estimation of survival and detection probabilities for multiple tagged salmon stocks with nested migration routes, using a large-scale telemetry array. *Mar. Freshw. Res.*, **60**:12 (2009), 1231–43.

Melnychuk, M. C., D. W. Welch, C. J. Walters, and V. Christensen, Riverine and early ocean migration and mortality patterns of juvenile steelhead trout (*Oncorhynchus mykiss*) from the Cheakamus River, British Columbia. *Hydrobiologia*, **582** (2007), 55–65.

Menot, L., M. Sibuet, R. S. Carney, *et al.*, New perceptions of continental margin biodiversity. In A. D. McIntyre (ed.), *Life in the World's Oceans: Diversity, Distribution, and Abundance* (Oxford: Blackwell Publishing Ltd., 2010), pp. 79–101.

Mills, C. E. and S. H. D. Haddock, Key to the Ctenophora. In J. T. Carlton (ed.), *Light and Smith's Manual: Intertidal Invertebrates of the Central California Coast* (Berkeley, CA: University of California Press, 2007), pp. 189–99.

Mills, C. E., S. H. D. Haddock, C. W. Dunn, and P. R. Pugh, Key to the Siphonophora. In J. T. Carlton (ed.), *Light and Smith's Manual: Intertidal Invertebrates of the Central California Coast* (Berkeley, CA: University of California Press, 2007), pp. 150–66.

Miloslavich, P. and E. Klein (eds.), *Caribbean Marine Biodiversity: The Known and Unknown* (Lancaster, PA: DEStech Publications, 2005).

OBIS, OBIS: explore data on locations of marine animals and plants. 2010. Available from: www.iobis.org.

Opdal, A. F., O. R. Godo, O. A. Bergstad, and O. Fiksen, Distribution, identity, and possible processes sustaining meso- and bathypelagic scattering layers on the northern Mid-Atlantic Ridge. *Deep Sea Res. II*, **55**:1–2 (2008), 45–58.

Pages, F., P. Flood, and M. Youngbluth, Gelatinous zooplankton net-collected in the Gulf of Maine and adjacent submarine canyons: new species, new family (Jeanbouilloniidae), taxonomic remarks and some parasites. *Sci. Mar.*, **70**:3 (2006), 363–79.

Payne, J., K. Andrews, C. Chittenden, *et al.*, Tracking fish movements and survival on the Northeast Pacific Shelf. In A. D. McIntyre (ed.), *Life in the World's Oceans: Diversity, Distribution, and Abundance* (Oxford: Blackwell Publishing Ltd., 2010), pp. 269–90.

Peckham, S. H., D. M. Diaz, A. Walli, *et al.*, Small-scale fisheries bycatch jeopardizes endangered Pacific loggerhead turtles. *PLoS ONE*, **2**:10 (2007), e1041.

Pimm, S. L. and P. Raven, Extinction by numbers. *Nature*, **403**:6772 (2000), 843–5.

Pitcher, C. R., P. Lawton, N. Ellis, *et al.*, Exploring the role of environmental variables in shaping patterns of biodiversity composition in seabed assemblages (unpublished manuscript).

Ramirez-Llodra, E., J. B. Company, F. Sarda, and G. Rotllant, Megabenthic diversity patterns and community structure of the Blanes submarine canyon and adjacent slope in the Northwestern Mediterranean: a human overprint? *Mar. Ecol.*, **32** (2010), 167–82.

Rigby, P. R., K. Iken, and Y. Shirayama, *Sampling Biodiversity in Coastal Communities: NaGIA Protocols for Seagrass and Macroalgal Habitats*, (Japan: Kyoto University Press, 2007).

Robison, B. H., The coevolution of undersea vehicles and deep-sea research. *Mar. Technol. Soc. J.*, **33**:4 (1999), 65–73.

Schindel, D. E. and S. E. Miller, DNA barcoding a useful tool for taxonomists. *Nature*, **435**:7038 (2005), 17.

Scholin, C., G. Doucette, S. Jensen, *et al.*, Remote detection of marine microbes, small invertebrates, harmful algae, and biotoxins using the environmental sample processor (ESP). *Oceanography*, **22**:2 (2009), 158–67.

Shaffer, S. A., Y. Tremblay, H. Weimerskirch, *et al.*, Migratory shearwaters integrate oceanic resources across the Pacific Ocean in an endless summer. *Proc. Natl. Acad. Sci. USA*, **103**:34 (2006), 12799–802.

Sirovich, L., M. Y. Stoeckle, and Y. Zhang, A scalable method for analysis and display of DNA sequences. *PLoS ONE*, **4**:10 (2009).

Snelgrove, P. V. R. and C. A. Butman, Animal-sediment relationships revisited: cause versus effect. *Oceanogr. Mar. Biol. Annu. Rev.* (1994), 111–78.

Sogin, M. L., H. G. Morrison, J. A. Huber, *et al.*, Microbial diversity in the deep sea and the underexplored "rare biosphere." *Proc. Natl. Acad. Sci. USA*, **103** (2006), 12115–20.

Solow, A. R., Inferring extinction from a sighting record. *Math. Biosci.*, **195**:1 (2005), 47–55.

Solow, A. R. and W. Smith, Missing and presumed lost: Extinction in the ocean and its inference (unpublished manuscript).

Teo, S. L. H., A. Boustany, S. Blackwell, *et al.*, Validation of geolocation estimates based on light level and sea surface temperature from electronic tags. *Mar. Ecol. Prog. Ser.*, **283** (2004), 81–98.

Teo, S. L. H., R. M. Kudela, A. Rais, *et al.*, Estimating chlorophyll profiles from electronic tags deployed on pelagic animals. *Aquat. Biol.*, **5**:2 (2009), 195–207.

Tittensor, D., A. R. Baco, P. E. Brewin, *et al.*, Predicting global habitat suitability for stony corals on seamounts. *J. Biogeogr.*, **36**:6 (2009), 1111–28.

Tittensor, D. P., F. Micheli, M. Nystrom, and B. Worm, Human impacts on the species-area relationship in reef fish assemblages. *Ecol. Lett.*, **10** (2007), 760–72.

Tolkien, J. R. R., *The Book of Lost Tales Vol. 1* (New York: Ballantine Books, 1992).

TOPP, Laysans come home to Guadalupe. 2007. Available from: www.topp.org/blog/laysans_come_home_guadalupe.

TOPP, Tagging of Pacific Predators. 2010; Available from: http://www.topp.org/species/bluefin_tuna.

TOPP, Tagging of Pacific predators: elephant seals. 2010. Available from: http://www.topp.org/species/elephant_seals.

Tsontos, V. M. and D. A. Kiefer, The Gulf of Maine biogeographical information system project: developing a spatial data management framework in support of OBIS. *Oceanol. Acta*, **25**:5 (2002), 199–206.

Twain, M., *Chapters from My Autobiography* (North American Review, No. DCXVII, 1907).

Vanden Berghe, E., K. I. Stocks, and J. F. Grassle, Data integration: the Ocean Biogeographic Information System. In A. D. McIntyre (ed.), *Life in the World's Oceans: Diversity, Distribution, and Abundance* (Oxford: Blackwell Publishing Ltd., 2010), pp. 333–53.

Vecchione, M., O. A. Bergstad, I. Byrkjedal, *et al.*, Biodiversity patterns and processes on the Mid-Atlantic Ridge. In A. D. McIntyre (ed.), *Life in the World's Oceans: Diversity, Distribution, and Abundance* (Oxford: Blackwell Publishing Ltd, 2010), pp. 103–21.

Walters, C. and S. Martell, *Fisheries Ecology and Management* (Princeton, NJ: Princeton University Press, 2004).

Welch, D. W., M. C. Melnychuk, E. R. Rechisky, *et al.*, Freshwater and marine migration and survival of endangered Cultus Lake sockeye salmon (*Oncorhynchus nerka*) smolts using POST, a large-scale acoustic telemetry array. *Can. J. Fish. Aquat. Sci.*, **66**:5 (2009), 736–50.

Welch, D. W., E. L. Rechisky, M. C. Melnychuk, *et al.*, Survival of migrating salmon smolts in large rivers with and without dams. *PLoS Biol.*, **6**:12 (2008), 2940.

Welch, D. W., S. Turo, and S. D. Batten, Large-scale marine and freshwater movements of white sturgeon. *Trans. Am. Fish. Soc.*, **135**:2 (2006), 386–9.

Welch, D. W., B. R. Ward, and S. D. Batten, Early ocean survival and marine movements of hatchery and wild steelhead trout (*Oncorhynchus mykiss*) determined by an acoustic array: Queen Charlotte Strait, British Columbia. *Deep Sea Res. II*, **51**:6–9 (2004), 897–909.

Weng, K. C., P. C. Castilho, J. M. Morrissette, *et al.*, Satellite tagging and cardiac physiology reveal niche expansion in salmon sharks. *Science*, **310**:5745 (2005), 104–6.

Wenneek, T. L., T. Falkenhaug, and O. A. Bergstad, Strategies, methods, and technologies adopted on the R.V. G.O. Sars MAR-ECO expedition to the Mid-Atlantic Ridge in 2004. *Deep Sea Res. II*, **55**:1–2 (2008), 6–28.

Wiebe, P. H., A. W. Morton, A. M. Bradley, *et al.*, New development in the MOCNESS, an apparatus for sampling zooplankton and micronekton. *Mar. Biol.*, **87**:3 (1985), 313–23.

Wiebe, P. H., A. Morton, L. P. Madin, *et al.*, Deep-sea holozooplankton species diversity in the Sargasso Sea, Northwestern Atlantic Ocean. *Deep Sea Res. II* (in press).

Witman, J. D., R. J. Etter, and F. Smith, The relationship between regional and local species diversity in marine benthic communities: a global perspective. *Proc. Natl. Acad. Sci. USA*, **101**:44 (2004), 15664–9.

Worm, B., H. K. Lotze, I. Jonsen, and C. Muir, The future of marine animal populations. In A. D. McIntyre (ed.), *Life in the World's Oceans: Diversity, Distribution, and Abundance* (Oxford: Blackwell Publishing Ltd., 2010), pp. 315–30.

Yoerger, D. R., A. M. Bradley, M. Jakuba, *et al.*, Autonomous and remotely operated vehicle technology for hydrothermal vent discovery, exploration, and sampling. *Oceanography*, **20**:1 (2007), 152–61.

5

Around the ocean rim

Coral reefs, rocky shorelines, and muddy bottoms add to the potentially millions of species in the ocean, but the range of habitats (Figure 5.1) from spectacular, species-rich coral reefs to species-poor intertidal pools adds them unevenly. Exploration of the ocean began easily with flipping over rocks on the seashore to see what hid beneath or snorkeling on a reef teeming with colorful fishes. But this strategy explores only a portion of the intersection of sea and humanity, missing many areas that support commercial fisheries and myriad ocean life.

Three field projects of the Census focused specifically on the coastal environment. The *Natural Geography in Shore Areas*, NaGISA, inventoried and monitored biodiversity in a narrow depth band of less than 20 meters wide around the world. They emphasized areas covered by *macroalgae*, or kelps and seaweeds, in rocky intertidal and subtidal environments and seagrass-covered, soft-bottom areas. The *Census of Coral Reef Ecosystems* project, CReefs, brought together knowledge on coral reef biodiversity and added new standardized approaches to taxonomy and field sampling that reduce the unknowns. Importantly, they developed a new blueprint for assessment of coral reef diversity for science and management applications. The *Gulf of Maine Area* project, GoMA, is the only Census project focused on managing a specific ecosystem – the Gulf of Maine. They showed how to integrate biodiversity knowledge in ecosystem-based management of ocean life. The *Future of Marine Animal Populations* (FMAP) and *History of Marine Animal Populations* (HMAP) intersect with these environments, but deal with the changing ocean that Chapter 9 analyzes. Other Census projects cross the ocean rim, but are dealt with in other chapters.

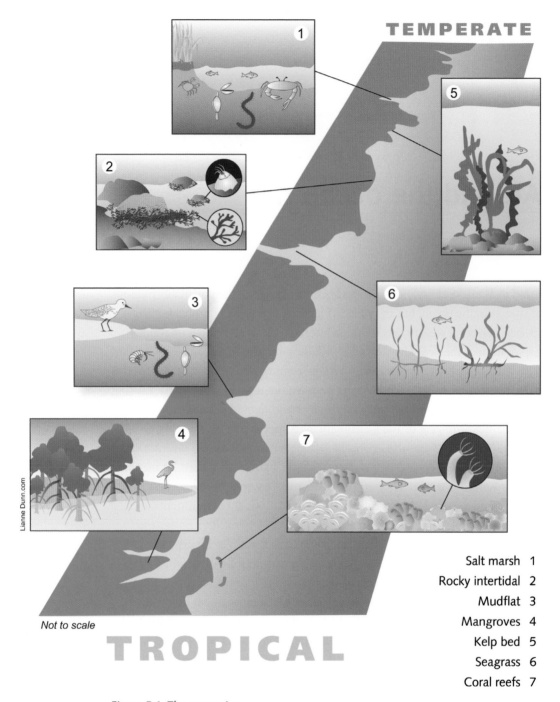

Figure 5.1 The ocean rim.
The drawing illustrates different coastal habitats studied by the Census from the tropics near the equator to temperate environments at mid-latitudes. Each habitat has unique characteristics and its own suite of species.

Along the shoreline

NaGISA focused on intertidal and shallow subtidal environments because their three-dimensional structure contributes to diversity more than adjacent sediments. The characteristics and species within these habitats vary from one region to another, but they span the poles to the tropics and are ideal for global-scale comparison.

To create their global dataset, NaGISA developed standardized sampling protocols for rocky intertidal and seagrass environments (Figure 5.2). Then, engaging school children, local residents, and scientists, NaGISA spanned 245 locations, sampling some repeatedly and thus establishing long-term monitoring locations. NaGISA initiated monitoring beyond coral reefs. The 80% decline in coral reefs may change adjacent rocky intertidal and subtidal life too, because some of their species are shared.

International workshops compiled taxonomic guides for macroalgae and key invertebrate groups, contributing to new taxonomic guides on echinoderms, hermit crabs, and seagrasses. NaGISA standardized taxonomy to ensure consistent names to a degree rarely achieved. Though a few previous studies used standard methodologies across a span of latitudes and others addressed inconsistent sampling, uniform global datasets remain rare and valuable.

NaGISA delivered to OBIS 37,000 records of specimens of 2,500 different taxa. These records include discoveries of species as small as inconspicuous crustaceans and as big as the highly visible 3-meter-long "golden V" kelp *Aureophycus aleuticus* from Alaska's Aleutian Islands (Figure 1.5c). NaGISA specimens also extend geographic ranges for previously known species along the coastal rim. Scientists learned that along the Argentinean shore, the invasive brown algae, *Undaria pinnatifida*, replaces native algae and attracts its own set of species during the winter. After an inactive winter, an offshore bank off Maine becomes a hot spot of abundant krill crustaceans that feed whales and seabirds.

Regional knowledge

Within individual countries, NaGISA studies built on established programs and knowledge. The European Union's *Marine Biodiversity and Ecosystem Functioning* (MarBEF) is bringing together data on the well-studied European Atlantic. The Census' NaGISA adds broad-scale analysis of a standardized

(a)

(b)

Figure 5.2 Tools for sampling life on the ocean rim.

The collage shows some tools that Census scientists used to sample coastal waters.

(a) Census researchers on an Antarctic ice floe tag seals with the newest generation of biologgers.

(b) Laying out PVC quadrats that define intertidal plots to be sampled in the Gulf of Maine, the late Robin Rigby, shown in the photograph, coordinated sampling.

(c) A light attracts animals for examination in an aquarium on Lizard Island Reef, Australia.

(d) A scuba diver deploys Autonomous Reef Monitoring Structures (ARMS), "condos" made of PVC, that mimic a reef. Small holes and gaps in the plastic ARMS provide homes for corals, crabs, and mollusks. Metal stakes secure the ARMS to a reef for a year or two until scientists retrieve the condos to see what types of life colonized them.

dataset that is only beginning, but early discoveries include higher abundances of individuals of rare species in more variable rocky shoreline areas. Within the Caribbean, the coordinated sampling effort has shown a general diversity decrease from west to east. In the northwest Atlantic each region has its characteristics and patterns. Oceanographic discontinuities and breaking waves define 13 rocky intertidal regions along the Pacific coast of North America.

Even strong Japanese science has not fully explored nearshore diversity in the western Pacific. The adjacent coral triangle region of the southwestern Pacific near Indonesia, Malaysia, and the Philippines is a hot spot of diversity awaiting exploration that spills over into the Japanese species pool.

In some NaGISA study areas, data are few or absent, such as the islands of India where three-quarters of the fauna is unknown. The approximately 12,000 species that live where the Indian and Atlantic Oceans meet represent 6% of all known marine coastal species, but other areas of African shorelines lack South Africa's scientific capacity and taxonomic emphasis, leaving marine life to be explored.

The southern hemisphere ocean is generally poorly sampled. For example, 540 new taxa have been described just from Brazilian seagrass beds and scientists recently discovered 50 new species in the fjords of southern Chile. Marine life in the southern hemisphere awaits discovery.

Break points

Because salinity, nutrients, seafloor composition, and water temperature often gently transition from one ecoregion to the next, so do suites of faunas and floras. Transition regions are hot spots of elevated diversity. In the Gulf of Maine, oceanographic conditions and tidal exposure define the habitats for over 800 species of invertebrates. Steep-sided Cashes Ledge in the Gulf of Maine, a feature that protrudes above the seabed into shallow depths, creates another species hot spot that may resemble historical rocky subtidal ecosystems in the Gulf.

Sharp changes of water masses also create diversity hot spots. The flora along the African coastline abruptly changes from tropical Indian Ocean species to *temperate*, or middle-latitude, species just south of the Mozambique border. The subtropical Kuroshio and subpolar Oyashio currents create a major faunal break on the coastline of Japan. The high biomass, large individuals, and low biodiversity fauna in the cool, nutrient-rich southbound Oyashio breaks sharply to the high biodiversity, but low biomass in the warm, northward-flowing Kuroshio. From east to west along the Alaskan Aleutian Archipelago, *Nereocystis luetkeana* gives way to a primary canopy-forming kelp, *Alaria fistulosa*.

Break points, such as the Wallace line that separates Asia and Australia, define the edges of ecoregions. Regional hot spots of diversity are found where many species share a relatively short stretch of coastline. Along the coast, a single sample may contain few species, but moving a short way adds more. If we do not know where and why these breaks and hot spots exist, we will assess biodiversity inaccurately.

From south to north

Latitudinal patterns of biodiversity along a coast likely resemble the familiar pattern on land where diversity in the tropics is high, but declines toward the poles. Oceanic mass and capacity to absorb and hold heat are great, dampening temperature differences compared to those on land, but similar latitudinal patterns occur in the ocean likely because of other factors.

Amid long-established declining diversity along Atlantic coasts at higher latitude, exploration has only recently filled in the details of hot spots like Cobscook Bay at the Canada–US Atlantic border. Along the Chilean coast, invertebrate diversity decreases from 18° to 40–45° South, before increasing again farther south, likely because sub-Antarctic fauna contribute to the species pool.

FMAP built on this work by analyzing multiple datasets. Moving north or south from the great coral diversity hot spot between Malaysia and Papua New Guinea, coral diversity declines. Among the few coral species in the Atlantic the pattern is similar, but less striking. Mangroves and seagrasses exhibit similar patterns, except with seagrass, diversity hot spots occur at 30° to the north and south of the equator. Pinniped seals, sea lions, and walruses are outliers that are most diverse at the poles, and the diversity of coastal cephalopods declines both north and south from a mid-latitude hot spot in the western Pacific. In summary, no "one size fits all" for latitudinal patterns of biodiversity. Although these patterns reflect evolutionary histories, a Census study of oceanic predators affirms key roles for temperature and oxygen, and their new analysis also adds chlorophyll, a measure of phytoplankton abundance, for some coastal groups. Despite less sampling of the southern hemisphere, the patterns there typically mirror those in the north.

Other data support a latitudinal pattern. The better known Antarctic macroalgal diversity declines from 62 to 68° South, and Arctic nearshore environments are less diverse than temperate areas. The catastrophic effects of glaciation during a mere blink for evolution some 10–15,000 years ago complicate Arctic patterns.

Ocean "rainforests"

The rich array of invertebrates and colorful fishes on coral reefs (Figure 5.3) signal species-rich environments. The coral requirements for clear, shallow waters and water temperatures above 20 °C limit them to less than 0.1% of Earth's surface and less than 0.2% of ocean area. But their complex, three-dimensional structure packs many species in a small area and contributes much to global biodiversity. Reef-building corals themselves number less than 1,000 species, but small invertebrates hiding cryptically in the cracks and crevices of living and dead coral comprise much of reef diversity. Marine ecologists believe that coral reefs support more species per unit area than any marine habitat; indeed more than one-third of all known marine species live on coral reefs. Only the vast deep sea rivals coral reefs in species, but more sparsely and over a much larger area.

Reef areas seem few and shallow, so it is surprising that estimates of species number are so tentative in a seemingly accessible environment. But many reefs lie along the coasts of developing nations with limited science infrastructure, or in remote places like mid-Pacific islands thousands of kilometers from any major seaport or airport, so many are poorly sampled. Finally, because the bulk of their diversity is hidden within the cracks and crevices of corals, it is difficult to sample and count.

How to census reef diversity? Beginning in 2005, the CReefs project developed a strategy to advance with every sample collected and slowly reduce the imprecision. Their strategy takes two approaches: genetic tools and standardized sampling for broad-scale comparisons.

Genetics and barcoding coral reef life highlights the strengths and weaknesses of these approaches and how they improve diversity estimates. Barcoding the COI gene does not work for corals or some crustaceans.

When CReefs sampled the central Pacific reefs of the Northern Line Islands the species were so poorly known that there was not a single match in *GenBank*, the global "phonebook" of known genetic sequences. This is like searching a well-illustrated field guide and finding none of the species you've collected. Because many reef species are unknown, CReefs proposes naming unknown ones by their barcodes, perhaps until taxonomists write traditional morphological descriptions to match the barcode names. CReefs has embraced the strategy of building up that "phonebook" because barcoding works quite well for many types of reef organisms.

CReefs' holy grail of identifying diversity of coral reefs applies mass sequencing "environmental genomics" that ICoMM developed for microbes to multicellular organisms. While the "biodiversity in a blender" appeals, its adaptation to multicellular organisms remains a challenge that will take time and effort to resolve. Some multicellular organisms produce compounds that interfere with DNA amplification, a key step in mass sequencing. Multicellular organisms vary in size and thus contain very different amounts of DNA per individual. The genetic signal from one large crab could swamp the signal from many hundreds of small crustaceans and mask the abundance of different species. Determining presence versus absence will be much easier than assessing abundance. Finally, because some DNA primers needed to determine genetic sequences don't work well in some groups of organisms, sequencing requires multiple primers. The "magic blender" will remain a potential technology until these problems are resolved.

To compare coral reefs, CReefs developed two standard sampling protocols to address the challenge of collecting representative samples, eliminating the problem that reef divers typically collect samples with different efficiencies and scope. First, CReefs collected similarly sized heads of dead *Pocillopora* coral to capture the cryptic life inside dead corals. Second, because this first strategy misses organisms that reside within live corals and other species, CReefs developed *Autonomous Reef Monitoring Structures* (ARMS) modified from an earlier design (Figure 5.2d). A series of PVC plates with spacers create an artificial structure on which reef organisms can settle and grow. These "reef condos" are anchored to different reefs and then retrieved after a set period of time to show what has moved in. Just as polling the residents of a Miami Beach condominium doesn't give a full picture of all residents of Miami, the reef condos target only those individuals that like this particular habitat. But ARMS expedite comparisons of reefs anywhere in the world, and some data suggest ARMS represent overall reef diversity.

(a)

Figure 5.3 Diverse ocean rim habitats.
From the tropics to the poles and from the intertidal to the subtidal, the ocean rim provides a variety of habitats.

(a) In shallow water, the sea star *Solaster*, the sea anemone *Tealia*, and sea urchins live on rocks in the rocky seashore of eastern Canada. Because these species mostly live in shallow water on rocky bottoms, scientists can study them easily.

(b) A coral reef, the Banco Chinchorro in Mexico, is home for the blue chromis, *Chromis cyanea*, and other fish. Reefs are rich in species.

(c) In sand, a giant guitarfish, *Rhynchobatus djiddensis*, also known as white-spotted guitarfish, whitespot ray, sandshark, and whitespot shovelnose ray, lies hidden off Lizard Island, Queensland, Australia.

(d) A rocky outcrop in shallow water off eastern Canada protects an American lobster, *Homarus americanus*, and provides home for attached invertebrates. Lobster now comprises the most important fishery in Eastern Canada and the northeast United States.

(b)

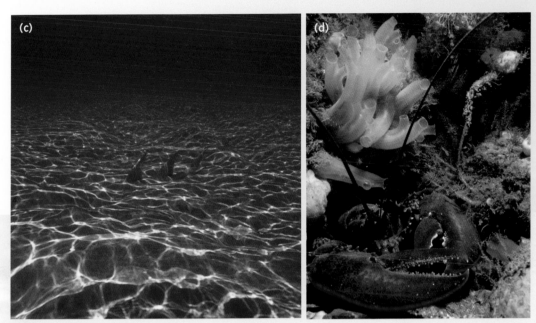

(e) Seaweed grows on seafloor exposed at low tide around Knight Island in Prince William Sound, Alaska. When low tide exposes intertidal life, Census scientists need only rubber boots to access their sample plots.

Clearly, a full assessment of life on reefs will also require divers and video to assess large and mobile organisms because "reef condos" and *Pocillopora* heads assess only organisms small enough to move in. And molecular tools specialize in differentiating similar-looking species. To capture the diversity of coral reef sediments and other reef environments, scientists must develop other standard protocols.

With these new tools CReefs took intriguing glimpses into the unknown diversity of coral reefs. They collected near Tahiti and four other reefs within a 1,000 kilometer span of the isolated Northern Line Islands. They examined 22 small dead *Pocillopora* heads and found 403 usable sequences from crustaceans. They discovered 135 potential crustacean species. Rarity was the rule; 44% of all the crustaceans were sampled just once, a pattern CReefs also found in areas such as Australia. Just 22 coral heads contained about 30% as many *brachyuran* (true) crabs as ever found in the most thoroughly sampled region of the global ocean, the European seas. Cruises and shore-based expeditions discovered about 100 new species from Hawaii and another 500 from Australia alone, with many more expected from sampling the Central Pacific. Like many projects, CReefs conducted workshops on molecular analyses and sampling protocols and they submitted more than 400,000 data records to OBIS.

The public, and scientists too, have paid less attention to coral reefs than to their terrestrial counterparts, tropical rainforests. Like rainforests, coral reefs are seriously threatened. Studies on coral reefs have focused on corals and fishes rather than the full array of reef diversity. This discrepancy may reflect the challenges of measuring reef diversity. Conservationists believe that the 60% of reefs degraded or gone are taking a significant chunk of their estimated 35% of global marine biodiversity with them. The loss of these reefs has other costs because the economic value of reefs is estimated at 30 billion US dollars annually. By the year 2000, humans had settled near more than 75% of the global reefs, a 25% increase since 1950.

Warming temperatures, increased pollution, ocean *acidification* (more acidic oceans caused by additional carbon dioxide), and overfishing all change food webs and stress reefs globally. These stresses cause *coral bleaching* that drives out photosynthetic cells that live within their tissue, eventually leaving bleached, dead coral. As the loss of corals eliminates many species that rely on them, the coral community typically begins to resemble a rocky shoreline community. Kelps and seaweeds replace the coral, creating habitats for other species, but creating a different and less diverse community.

Surprisingly and contrary to earlier evidence of impacts on fishes and corals, CReefs found that humans did not diminish crustacean diversity in the Northern Line Islands. These remote reefs are relatively pristine compared to many reefs near humans, and moderately impacted reefs sustain high levels of biodiversity compared to the species losses on severely impacted reefs like many in the Caribbean, where many invertebrates are declining.

The most comprehensive dataset ever assembled on reef fishes shows that the productivity of reefs grows with biodiversity of fishes, making the most diverse reefs the most productive ones. This illustrates *functional redundancy* in ecology, where more diverse environments may withstand disturbance because multiple species fill similar key roles in food webs and other functions. A species that plays the same ecological role as another insures against the loss of one. These comprehensive data suggest not that diverse reefs are invulnerable, but rather that diversity helps mitigate the pervasive human footprint on reefs.

Cooler waters in the Gulf of Maine

The Gulf of Maine from Cape Cod north to the Nova Scotian shelf and Bay of Fundy provides a cooler ocean world. The Census project GoMA chose these relatively well sampled and less diverse waters to make existing information accessible for ecosystem-based management.

Though the beaches and rocky coast of Maine and Atlantic Canada may seem pristine, humans have been there for 6,000 to 8,500 years. Early hunter-gatherers had little impact, but evidence from archaeological *middens*, or ancient garbage dumps, shows changed food webs, including cod, 3,500 years ago. European colonization in the 1700s transformed parts of the environment by loading nutrients, altering seabed habitat, and over-fishing. Europeans significantly reduced cod on the Scotian shelf by 1859 and overexploited most vertebrates. They hunted three species of mammals and six species of birds to extinction. In the early 1900s, mechanized and more efficient fishing technologies depleted coastal and Georges Bank cod. By 2007, cod landings in the Gulf of Maine system fell to 5–6% of their 1861 levels. Multiple effects cascaded through the ecosystem. Fishing still affects the Gulf of Maine ecosystem despite new restrictions that recognize the importance of habitat, vulnerable species, and biodiversity.

Against this backdrop in the Gulf of Maine, GoMA assembled biodiversity knowns and unknowns, and set a research strategy. They designed a system for available information, supported field efforts in key areas, and moved biodiversity knowledge into ecosystem-based management.

Over four decades, fish surveys by US and Canadian fisheries agencies created a goldmine of knowledge about the Gulf of Maine. GoMA analyses show that fish diversity is lowest on the shelf and in basins around the Gulf of Maine and highest on Georges Bank and on the upper continental slope. Because fishes associate with specific types of seabed habitat, managers may be able to extrapolate where Gulf of Maine biodiversity hot spots are likely. This is just the sort of information needed for ecosystem-based management.

GoMA helped develop the concept of a *Discovery Corridor,* to colocate biodiversity research where it is needed and most productive. The Discovery Corridor in the Gulf of Maine extends in a triangle that fans out from the intertidal region near the Canada–US border across the continental shelf and to the base of the continental slope. This triangle encompasses a variety of habitats from the rocky shoreline to areas that contain the greatest known regional abundances of deep-water corals, such as sea corn, *Primnoa resedaeformis,* and bubble gum coral, *Paragorgia arborea.* Genetic analyses suggest species of coral and other invertebrates previously unknown from Canadian waters and potentially new. Because scientists know little about the deep water of the Discovery Corridor and adjacent seamounts, GoMA assembled experts to gather available data and plan geographic and taxonomic strategies to tackle the unknowns.

In the Gulf of Maine, like elsewhere, scientists know the least about microbes which likely add the greatest diversity. Some 696 known species of phytoplankton, plus many unknown microbes, need study. GoMA microbial experts extrapolated upwards from estimated microbial abundance and analysis of abundant taxa to estimate as many as 10,000 types of phytoplankton and 400,000 types of bacterioplankton. Bacterioplankton in the Gulf represent about 20% of the known global diversity total, hinting that the microbial diversity in the Gulf of Maine may be especially rich. High-throughput analysis offers the potential for monitoring microbes as harbingers of ecosystem health in the Gulf of Maine and elsewhere.

A GoMA project illustrates the human bias in how we view, sample, and even manage the ocean. Figure 5.4 compares size of adult stages of different species and reinforces the common-sense view that the smaller the organism, the less we know. Whales and fishes are almost fully known, invertebrates

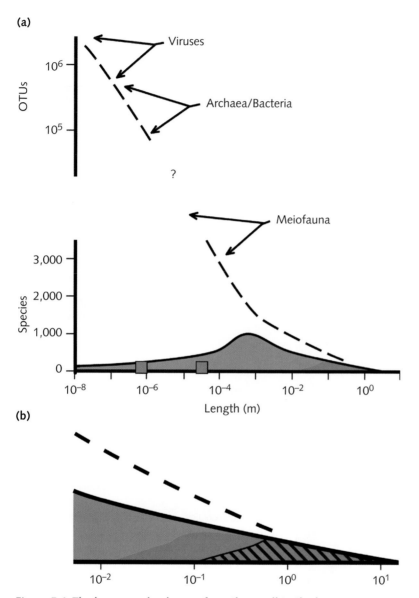

Figure 5.4 The known and unknown from the small to the large.
Although the polychaetes, mollusks, and crustaceans near 1 millimeter in size are
the best known among the small creatures, many species of that size are still unknown.
Still smaller life, teeming forms that can only be identified by genetic analysis, are easily
overlooked altogether. There may be millions of viruses, archaea, and bacteria. In the
upper figure **(a)** scientists know some groups well (blue), monitor a small subset of this
biodiversity (orange), and attempt to manage an even smaller subset (orange hatched).
The lower figure **(b)** enlarges the part of the upper figure (a) that illustrates the
larger organisms such as fishes and marine mammals, most of which are known.

Figure 5.5 Life on the ocean rim.
The species that inhabit the ocean rim range from kelps to sharks, from the tropics to mid-latitudes to the poles described in Chapter 6.
(a) A cerianthid anemone off eastern Canada catches small particles from water flowing past its tentacles.
(b) A white-topped coral crab from a dead coral head off Heron Island, Australia.
(c) A whale shark, *Rhincodon typus*, photographed on Ningaloo Reef, Australia.
(d) An Arctic *Solaster* sea star photographed in Alaskan shallow waters.
(e) *Nereocystis*, the bull kelp, lives in shallow water off the Pacific coast of North America.
(f) An octopus collected at a depth of about 10 meters off Lizard Island, Australia with ARMS (Autonomous Reef Monitoring Structure).

are partly, but unevenly, known, and microbes are almost totally unknown. To illustrate these points, the cycling of food and nutrients in the Gulf of Maine and elsewhere is driven largely by microbes within a "*microbial loop*," yet many are unknown. Compare lists of known species from the Gulf of Maine to well-known European waters and assume proportions of phyla are similar. The conclusion: small creatures from nematodes to small crustaceans are scarcely sampled and thousands of invertebrate species are unknown. Cruises over the last decade already discovered many new species, including cryptic forms that are virtually impossible to distinguish based on morphology alone.

GoMA discovered schools of herring over 40 kilometers wide move into shallow water. Large-scale acoustic imaging shows that schools of this key commercial species form and disperse in minutes to hours. The herring move quickly from the sides of Georges Bank as they form spawning shoals at the top of the bank near sunset. A disorganized mass of herring assemble into a highly organized school that follows a few leaders. As the fish assemble, they stir a convergence wave in the water that propagates outward at 10–20 kilometers per hour. In about 40 minutes, hundreds of millions of fish across tens of kilometers of ocean synchronize. Imagine assembling and moving the entire human population of the United States in synchrony in 40 minutes!

The US National Marine Sanctuaries include Stellwagen Bank, a 2,100-square-kilometer area on the southwest corner of the Gulf of Maine that has been fished more than 400 years. Most activities are permitted on the Bank, as in most parts of 200 protected areas in the Gulf of Maine. Like the Discovery Corridor, it attracts researchers studying fishes to plankton to seabirds and marine mammals, using varied tools. For example, the towed HabCam camera makes automated species identifications, and when paired with broader-scale maps adds useful information for managers. These different types of information work in statistical analyses such as *random forests models* adapted to the sea. These analyses now determine how gradients in temperature, seafloor composition, and other environmental characters affect benthic species in the Gulf of Maine and elsewhere. This approach points to physical variables measured easily over broad areas as "surrogate" predictors of species, communities, and biodiversity patterns.

The incorporation of biodiversity knowledge into decision making increasingly guides ecosystem-based management and related management that integrate diverse information. Biodiversity and marine habitat mapping may be incorporated into ecosystem-based management from scales of genes to ecoregions. Because much-needed information is unknown, even for familiar regions like the Gulf of Maine, scientists must assemble and stretch the known into multiple applications. With much known about many species in many ecosystems, particularly those targeted by fisheries, bringing that knowledge together is a knowable next step.

The *Gulf of Maine Register of Marine Species* lists known species in the Gulf of Maine, including species thought to occur based on knowledge from adjacent regions. Once completed, this list will link to the *Encyclopedia of Life*, OBIS, and elsewhere. In November 2009, the Register listed 3,141 species, about a third were verified. An additional 821 species from various datasets and habitats increase their provisional list to 3,962 species.

New knowledge offers an opportunity to manage coastal environments better. Linking knowledge along the shoreline through regions, hot spots, and break points from north to south, will make management more productive. Exploration of the broad array of coastal life from warm coral reefs to the cool waters of the Gulf of Maine and beyond (Figure 5.5) will collectively propel ecosystem-based management for more sustainable ocean use.

BIBLIOGRAPHY

Abebe, E., R. E. Grizzle, D. Hope, and W. K. Thomas, Nematode diversity in the Gulf of Maine, USA, and a web-accessible, relational database. *J. Mar. Biol. Assoc. UK*, **84**:6 (2004), 1159–67.

Alexander, K. E., W. B. Leavenworth, J. Cournane, *et al.*, Gulf of Maine cod in 1861: historical analysis of fishery logbooks, with ecosystem implications. *Fish Fish.*, **10**:4 (2009), 428–49.

Amaral-Zettler, L., L. F. Artigas, J. Baross, *et al.*, A global census of marine microbes. In A. D. McIntyre (ed.), *Life in the World's Oceans: Diversity, Distribution, and Abundance* (Oxford: Blackwell Publishing Ltd., 2010), pp. 223–45.

Auster, P. J., R. B. Clark, and R. E. S. Reed, Chapter 3. Marine fishes. In T. Battista, R. Clark, S. Pittmann (eds.), *An Ecological Characterization of the Stellwagen Bank National Marine Sanctuary Region: Oceanographic, Biogeographic, and Contaminants Assessment* (Silver Spring, MD: NOAA, 2006).

Auster, P. J., K. Joy, and P. C. Valentine, Fish species and community distributions as proxies for seafloor habitat distributions: the Stellwagen Bank National Marine Sanctuary example (Northwest Atlantic, Gulf of Maine). *Environ. Biol. Fishes*, **60**:4 (2001), 331–46.

Auster, P. J. and J. Lindholm, The ecology of fishes on deep boulder reefs in the western Gulf of Maine. In *Diving for Science 2005, Proceedings of the American Academy of Underwater Science* (Groton, CT: Connecticut Sea Grant, 2005), pp. 89–107.

Auster, P. J. and N. L. Shackell, Marine protected areas for the temperate and boreal Northwest Atlantic: the potential for sustainable fisheries and conservation of biodiversity. *Northeast Nat.*, **7**:4 (2000), 419–34.

Awad, A. A., C. L. Griffiths, and J. K. Turpie, Distribution of South African marine benthic invertebrates applied to the selection of priority conservation areas. *Divers. Distrib.*, **8**:3 (2002), 129–45.

Benedetti-Cecchi, L., I. Bertocci, S. Vaselli, E. Maggi, and F. Bulleri, Neutrality and the response of rare species to environmental variance. *PLoS ONE*, **3**:7 (2008).

Blanchette, C. A., C. M. Miner, P. T. Raimondi, *et al.*, Biogeographical patterns of rocky intertidal communities along the Pacific coast of North America. *J. Biogeogr.*, **35**:9 (2008), 1593–607.

Bouchet, P., P. Lozouet, P. Maestrati, and V. Heros, Assessing the magnitude of species richness in tropical marine environments: exceptionally high numbers of molluscs at a New Caledonia site. *Biol. J. Linn. Soc.*, **75**:4 (2002), 421–36.

Bourque, B. J., *Twelve Thousand Years: American Indians in Maine* (Lincoln, NB: University of Nebraska Press, 2001).

Bourque, B. J., B. J. Johnson, and R. Steneck, Possible prehistoric fishing effects on coastal marine food webs in the Gulf of Maine. In T. C. Rick and J. Erlandson (eds.), *Human Impacts on Ancient Marine Ecosystems* (Berkeley, CA: University of California Press, 2008), pp. 165–85.

Branch, G. M., C. L. Griffiths, M. L. Branch, and L. E. Beckley, *Two Oceans: A Guide to the Marine Life of Southern Africa* (South Africa: Struik Publishers, 1994).

Brinton, E., M. D. Ohman, A. W. Townsend, M. D. Knight, and A. L. Bridgeman, *Euphausiids of the World Ocean* (Berlin: Springer Verlag, 2000).

Buzeta, M. I. and R. Singh, Identification of ecologically and biologically significant areas in the Bay of Fundy, Gulf of Maine. *Can. Tech. Rep. Fish. Aquat. Sci.*, **2788** (2008), 80 pp.

Buzeta, M. I., R. Singh, and S. Young-Lai, Identification of significant marine and coastal areas in the Bay of Fundy. *Can. Manu. Rep. Fish. Aquat. Sci.*, **2635** (2003), 177 pp.

Cairns, S., Species richness of recent Scleractina. *Atoll Res. Bull.*, **459** (1999), 1–46.

Cairns, S. D., Studies on western Atlantic Octocorallia (Gorgonacea: Primnoidae). Part 8: new records of Primnoidae from the New England and Corner Rise Seamounts. *Proc. Biol. Soc. Wash.*, **120**:3 (2007), 243–263.

Campbell, D. E., Evaluation and energy analysis of the Cobscook Bay ecosystem. *Northeast Nat.*, **11**: Special Issue 2 (2004), 355–424.

Claesson, S. and A. A. Rosenberg, *Stellwagen Bank Marine Historical Ecology: Final Report. Gulf of Maine Cod Project.* University of New Hampshire (2009).

Clark, R., J. Manning, B. Costa, *et al.*, Chapter 1. Physical and oceanographic setting. In T. Battista, R. Clark, S. Pittmann (eds.), *An Ecological Characterization of the Stellwagen Bank National Marine Sanctuary Region: Oceanographic, Biogeographic, and Contaminants Assessment* (Silver Spring, MD: NOAA, 2006), pp. 1–58.

Clarke, A. and J. A. Crame, Diversity, latitude and time: patterns in the shallow sea. In R. F. G. Ormond, J. D. Gage, and M. V. Angel (eds.), *Marine Biodiversity* (Cambridge: Cambridge University Press, 1997), pp. 122–45.

Cogan, C. B. and T. T. Noji, Marine classification, mapping, and biodiversity analysis. In B. J. Todd and H. G. Greene (eds.), *Mapping the Seafloor for Habitat Characterization* (St. John's, Newfoundland: Geological Association of Canada, 2007), pp. 129–39.

Cogan, C. B., B. J. Todd, P. Lawton, and T. T. Noji, The role of marine habitat mapping in ecosystem-based management. *ICES J. Mar. Sci.*, **66**:9 (2009), 2033–42.

Cogswell, A. T., E. L. R. Kenchington, C. G. Lirette, *et al.*, The current state of knowledge concerning the distribution of coral in the Maritime Provinces. *Can. Tech Rep. Fish. Aquat. Sci.*, **2855** (2009) 66 pp.

Conservation International, *Economic Values of Coral Reefs, Mangroves, and Seagrasses: A Global Compilation* (Arlington, VA: Center for Applied Biodiversity Science, 2008).

Couto, E. C. G., F. L. Da Silveira, and G. A. Rocha, Marine biodiversity in Brazil: the current status. *Gayana*, **67**:2 (2003), 327–40.

Curtis, T. P., W. T. Sloan, and J. W. Scannell, Estimating prokaryotic diversity and its limits. *Proc. Natl. Acad. Sci. USA*, **99**:16 (2002), 10494–9.

Dayton, P. K., S. F. Thrush, M. T. Agardy, and R. J. Hofman, Environmental effects of marine fishing. *Aquat. Conserv.: Mar. Freshwat. Ecosyst.*, **5**:3 (1995), 205–32.

De'ath, G., J. M. Lough, and K. E. Fabricius, Declining coral calcification on the Great Barrier Reef. *Science*, **323**:5910 (2009), 116–19.

Dunton, K. and S. Schonberg, The benthic faunal assemblage of the Boulder Patch kelp community. In J. C. Truett and S. R. Johnson (eds.), *The Natural History of an Arctic Oil Field* (California, USA: Academic Press, 2000), pp. 371–98.

EOL, Encyclopedia of Life. 2010. Available from: http://www.eol.org.

Fautin, D., P. Dalton, L. S. Incze, *et al.*, An overview of marine biodiversity in United States waters. *PLoS ONE* **5**:8 (2010), e11914, doi:10.1371/journal.pone.0011914.

Fernandez, M., E. Jaramillo, P. A. Marquet, *et al.*, Diversity, dynamics and biogeography of Chilean benthic nearshore ecosystems: an overview and guidelines for conservation. *Rev. Chil. Hist. Nat.*, **73**:4 (2000), 797–830.

Foote, K. G., Pilot census of marine life in the Gulf of Maine: contributions of technology. *Oceanol. Acta*, **25**:5 (2002), 213–8.

Frank, K. T., B. Petrie, J. S. Choi, and W. C. Leggett, Trophic cascades in a formerly cod-dominated ecosystem. *Science*, **308**:5728 (2005), 1621–3.

Frid, C., C. Hammer, R. Law, *et al.*, *Environmental Status of the European Seas* (Copenhagen: ICES. 2003).

Gallardo, V. A., The sublittoral macrofaunal benthos of the Antarctic Shelf. *Environ. Int.*, **13**:1 (1987), 71–81.

Gardner, T. A., I. M. Cote, J. A. Gill, A. Grant, and A. R. Watkinson, Long-term region-wide declines in Caribbean corals. *Science*, **301**:5635 (2003), 958–60.

Gavaris, S., Fisheries management planning and support for strategic and tactical decisions in an ecosystem approach context. *Fish. Res.*, **100**:1 (2009), 6–14.

Gerken, S., Two new *Cumella* (Crustacea: Cumacea: Nannastacidae) from the North Pacific, with a key to the North Pacific Cumella. *Zootaxa*, **2149** (2009), 50–61.

Gibbons, M. J., The taxonomic richness of South Africa's marine fauna: a crisis at hand. *S. Afr. J. Sci.*, **95**:1 (1999), 8–12.

Glynn, P. W., Coral-reef bleaching – ecological perspectives. *Coral Reefs*, **12**:1 (1993), 1–17.

Griffiths, C. L., Coastal marine biodiversity in East Africa. *Indian J. Mar. Sci.*, **34**:1 (2005), 35–41.

Hartel, K. E., C. P. Kenaley, J. K. Galbraith, and T. T. Sutton, Additional records of deep-sea fishes from off greater New England. *Northeast Nat.*, **15**:3 (2008), 317–34.

Häussermann, V. and G. Föresterra, eds, *Marine Benthic Fauna of Chilean Patagonia*, (Santiago: Nature in Focus, 2010).

Hebert, P. D. N., A. Cywinska, S. L. Ball, and J. R. DeWaard, Biological identifications through DNA barcodes. *Proc. R. Soc. Lond. B Biol. Sci.*, **270**:1512 (2003), 313–21.

Heip, C., H. Hummel, P. van Avesaath, *et al.*, *Marine Biodiversity and Ecosystem Functioning* (Dublin, Ireland: Printbase, 2009).

Hicks, G. R. F., Meiofauna associated with rocky shore algae. In P. G. Moore and R. Seed (eds.), *The Ecology of Rocky Coasts* (London: Hodder & Staughton, 1985), pp. 36–56.

Idjadi, J. A. and P. J. Edmunds, Scleractinian corals as facilitators for other invertebrates on a Caribbean reef. *Mar. Ecol. Prog. Ser.*, **319** (2006), 117–27.

Incze, L. S., P. Lawton, S. L. Ellis, and N. H. Wolff, Biodiversity knowledge and its application in the Gulf of Maine area. In A. D. McIntyre (ed.), *Life in the World's Oceans: Diversity, Distribution, and Abundance* (Oxford: Blackwell Publishing Ltd., 2010), pp. 43–63.

Jackson, J. B. C., Ecological extinction and evolution in the brave new ocean. *Proc. Natl. Acad. Sci. USA*, **105**:Suppl. 1 (2008), 11458–65.

Jackson, J. B. C., M. X. Kirby, W. H. Berger, *et al.*, Historical overfishing and the recent collapse of coastal ecosystems. *Science*, **293**:5530 (2001), 629–38.

Jereb, P. and C. F. E. Roper, *Cephalopods of the World. An Annotated and Illustrated Catalogue of Species Known to Date* (Rome: FAO, 2005).

Kawai, H., T. Hanyuda, M. Lindeberg, and S. C. Lindstrom, Morphology and molecular phylogeny of *Aureophycus aleuticus* gen. et sp nov (Laminariales, Phaeophyceae) from the Aleutian Islands. *J. Phycol.*, **44**:4 (2008), 1013–21.

Kerswell, A. P., Global biodiversity patterns of benthic marine algae. *Ecology*, **87**:10 (2006), 2479–88.

Kleypas, J., R. A. Feely, V. J. Fabry, *et al.*, *Impacts of Ocean Acidification on Coral Reefs and Other Marine Calcifiers: A Guide for Future Research* Washington, DC: NSF, NOAA, and the USGS (2006).

Knowlton, N., Coral reef coda: what can we hope for? In I. Cote and J. Reynolds (eds.), *Coral Reef Conservation* (Cambridge, UK: Cambridge University Press, 2006), pp. 538–49.

Knowlton, N., R. E. Brainard, R. Fisher, *et al.*, Coral reef biodiversity. In A. D. McIntyre (ed.), *Life in the World's Oceans: Diversity, Distribution, and Abundance* (Oxford: Blackwell Publishing Ltd., 2010), pp. 65–77.

Knowlton, N. and J. B. C. Jackson, Shifting baselines, local impacts, and global change on coral reefs. *PLoS Biol.*, **6**:2 (2008), e54.

Konar, B., K. Iken, G. Pohle, *et al.*, Surveying nearshore biodiversity. In A. D. McIntyre (ed.), *Life in the World's Oceans: Diversity, Distribution, and Abundance* (Oxford: Blackwell Publishing Ltd., 2010), p. 27.

Kuklinski, P. and D. K. A. Barnes, Structure of intertidal and subtidal assemblages in Arctic vs temperate boulder shores. *Pol. Polar Res.*, **29**:3 (2008), 203–18.

Li, W. K. W. and W. G. Harrison, Chlorophyll, bacteria and picophytoplankton in ecological provinces of the North Atlantic. *Deep Sea Res. II*, **48**:10 (2001), 2271–93.

Lotze, H. K., H. S. Lenihan, B. J. Bourque, *et al.*, Depletion, degradation, and recovery potential of estuaries and coastal seas. *Science*, **312**:5781 (2006), 1806–9.

Lotze, H. K. and I. Milewski, Two centuries of multiple human impacts and successive changes in a North Atlantic food web. *Ecol. Appl.*, **14**:5 (2004), 1428–47.

Makris, N. C., P. Ratilal, S. Jagannathan, *et al.*, Critical population density triggers rapid formation of vast oceanic fish shoals. *Science*, **323**:5922 (2009), 1734–7.

May, R. M., Tomorrow's taxonomy: collecting new species in the field will remain the rate-limiting step. *Philos. Trans. R. Soc. Lond. B Biol. Sci.*, **359**:1444 (2004), 733–4.

McLeod, K. L. and H. Leslie, *Ecosystem-based Management for the Oceans* (Washington, DC: Island Press, 2009).

Miller, K. A. and J. A. Estes, Western range extension for *Nereocystis luetkeana* in the North Pacific Ocean. *Bot. Mar.*, **32**:6 (1989), 535–8.

Miloslavich, P., J. M. Díaz, E. Klein, J. J. Alvavado, C. Díaz, et al., Marine biodiversity in the Caribbean: Regional estimates and distribution patterns. *PLoS ONE* **5**:8 (2010), e11916, doi:10.1371/journal.pone.0011916.

Moe, R. L. and T. E. Delaca, Occurrence of macroscopic algae along Antarctic Peninsula. *Antarct. J. US*, **11**:1 (1976), 20–4.

Moore, J. A., K. E. Hartel, J. E. Craddock, and J. K. Galbraith, An annotated list of deepwater fishes from off the New England region, with new area records. *Northeast Nat.*, **10**:2 (2003), 159–248.

Moore, J. A., M. Vecchione, B. B. Collette, R. Gibbons, and K. E. Hartel, Selected fauna of Bear Seamount (New England Seamount chain), and the presence of "natural invader" species. *Arch. Fish. Mar. Res.*, **51**:1–3 (2004), 241–50.

Mora, C., O. Aburto-Oropeza, A. A. Ayala Bocos, *et al.*, Human footprint on the linkage between biodiversity and ecosystem functioning in reef fishes (unpublished manuscript).

Murawski, S. A., R. Brown, H. L. Lai, P. J. Rago, and L. Hendrickson, Large-scale closed areas as a fishery-management tool in temperate marine systems: the Georges Bank experience. *Bull. Mar. Sci.*, **66**:3 (2000), 775–98.

Myers, R. A. and B. Worm, Rapid worldwide depletion of predatory fish communities. *Nature*, **423**:6937 (2003), 280–3.

Nishimura, S., *History of Japan Sea: Approach from Biogeography* (Tokyo: Tsukiji-shokan, 1974).

O'Boyle, R. and G. Jamieson, Observations on the implementation of ecosystem-based management: experiences on Canada's east and west coasts. *Fisheries Research*, **79**:1–2 (2006), 1–12.

O'Boyle, R. and T. Worcester, Eastern Scotian Shelf. In K. L. McLeod and H. Leslie (eds.), *Ecosystem-based Management for the Oceans* (Washington, DC: Island Press, 2009), pp. 253–67.

Pandolfi, J. M., R. H. Bradbury, E. Sala, *et al.*, Global trajectories of the long-term decline of coral reef ecosystems. *Science*, **301**:5635 (2003), 955–8.

Peters, J., B. De Baets, N. E. C. Verhoest, *et al.*, Random forests as a tool for ecohydrological distribution modelling. *Ecol. Model.*, **207**:2–4 (2007), 304–18.

Pitcher, C. R., P. Lawton, N. Ellis, *et al.*, Exploring the rare of environmental variables in shaping patterns of biodiversity composition in seabed assemblages (unpublished manuscript).

Pittmann, S., B. Cosat, C. Kot, *et al.*, Chapter 5. Cetacean distribution and diversity. In T. Battista, R. Clark and S. Pittmann (eds.), *An Ecological Characterization of the Stellwagen Bank National Marine Sanctuary Region: Oceanographic, Biogeographic, and Contaminants Assessment* (Silver Spring, Maryland: NOAA, 2006), pp. 264–324.

Pittmann, S. and F. Huetmann, Chapter 4. Seabird distribution and diversity. In T. Battista, R. Clark, S. Pittmann (eds.), *An Ecological Characterization of the Stellwagen Bank National Marine Sanctuary Region: Oceanographic, Biogeographic, and Contaminants Assessment* (Silver Spring, MD: NOAA, 2006), pp. 230–63.

Plaisance, L., J. Caley, R. E. Brainard, and N. Knowlton, The diversity of coral reefs: What could we be losing? (unpublished manuscript).

Plaisance, L., N. Knowlton, G. Paulay, and C. Meyer, Reef-associated crustacean fauna: biodiversity estimates using semi-quantitative sampling and DNA barcoding. *Coral Reefs*, **28**:4 (2009), 977–86.

Pollock, L. W., *A Practical Guide to the Marine Animals of Northeastern North America* (New Jersey: Rutgers University Press, 1998).

Rahayu, D. L. and A. J. Wahayudi (eds.), *Common Littoral Hermit Crabs of Indonesia* (Kyoto, Japan: Kyoto University Press, 2008).

Reaka-Kudla, M. L., The global biodiversity of coral reefs: a comparison with rain forests. In M. L. Reaka-Kudla, D. E. Wilson, and E. O. Wilson (eds.), *Biodiversity II: Understanding and Protecting Our Biological Resources* (Washington, DC: Joseph Henry Press, 1997), pp. 83–108.

Reaka-Kudla, M. L., Biodiversity of Caribbean coral reefs. In P. Miloslavich and E. Klein (eds.), *Caribbean Marine Biodiversity: The Known and the Unknown* (Lancaster, PA: DEStech Publications, 2005), pp. 259–76.

Rigby, P. R., K. Iken, and Y. Shirayama, *Sampling Biodiversity in Coastal Communities: NaGISA Protocols for Seagrass and Macroalgal Habitats* (Japan: Kyoto University Press, 2007).

Roberts, C. M., C. J. McClean, J. E. N. Veron, *et al.*, Marine biodiversity hotspots and conservation priorities for tropical reefs. *Science*, **295**:5558 (2002), 1280–4.

Rosenberg, A. A., W. J. Bolster, K. E. Alexander, *et al.*, The history of ocean resources: modeling cod biomass using historical records. *Front. Ecol. Environ.*, **3**:2 (2005), 84–90.

Roy, K., D. Jablonski, and J. W. Valentine, Dissecting latitudinal diversity gradients: functional groups and clades of marine bivalves. *Proc. R. Soc. Lond. B Biol. Sci.*, **267**:1440 (2000), 293–9.

Rutherford, S., S. D'Hondt, and W. Prell, Environmental controls on the geographic distribution of zooplankton diversity. *Nature*, **400**:6746 (1999), 749–53.

Sala, E. and N. Knowlton, Global marine biodiversity trends. *Annu. Rev. Environ. Resour.*, **31** (2006), 93–122.

Sandin, S. A., J. E. Smith, E. E. DeMartini, *et al.*, Baselines and degradation of coral reefs in the Northern Line Islands. *PLoS ONE*, **3**:2 (2008), e1548.

Saunders, G. W., Applying DNA barcoding to red macroalgae: a preliminary appraisal holds promise for future applications. *Philos. Trans. R. Soc. Lond. B Biol. Sci.*, **360**:1462 (2005), 1879–88.

Saunders, G. W., A DNA barcode examination of the red algal family Dumontiaceae in Canadian waters reveals substantial cryptic species diversity. 1. The foliose Dilsea-Neodilsea complex and Weeksia. *Can. J. Bot.*, **86**:7 (2008), 773–89.

Schipper, J., J. S. Chanson, F. Chiozza, *et al.*, The status of the world's land and marine mammals: diversity, threat, and knowledge. *Science*, **322**:5899 (2008), 225–30.

Schoch, G. C., B. A. Menge, G. Allison, *et al.*, Fifteen degrees of separation: latitudinal gradients of rocky intertidal biota along the California Current. *Limnol. Oceanogr.*, **51**:6 (2006), 2564–85.

Shearer, T. L. and M. A. Coffroth, Barcoding corals: limited by interspecific divergence, not intraspecific variation. *Molecul. Ecol. Res.*, **8**:2 (2008), 247–55.

Sherman, K., The large marine ecosystem concept – research and management strategy for living marine resources. *Ecol. Appl.*, **1**:4 (1991), 349–60.

Sherr, E. and B. Sherr, Marine microbes: an overview. In D. L. Kirchman (ed.), *Microbial Ecology of the Oceans* (New York, NY: Wiley-Liss, 2000), pp. 13–46.

Simpson, A. W. and L. Watling, An investigation of the cumulative impacts of shrimp trawling on mud-bottom fishing grounds in the Gulf of Maine: effects on habitat and macrofaunal community structure. *ICES J. Mar. Sci.*, **63**:9 (2006), 1616–30.

Steele, J. H., J. S. Collie, J. J. Bisagni, *et al.*, Balancing end-to-end budgets of the Georges Bank ecosystem. *Prog. Oceanogr.*, **74**:4 (2007), 423–48.

Steneck, R. S., J. Vavrinec, and A. V. Leland, Accelerating trophic-level dysfunction in kelp forest ecosystems of the western North Atlantic. *Ecosystems*, **7**:4 (2004), 323–32.

Stepanauskas, R. and M. E. Sieracki, Matching phylogeny and metabolism in the uncultured marine bacteria, one cell at a time. *Proc. Natl. Acad. Sci. USA*, **104**:21 (2007), 9052–7.

Stevick, P. T., L. S. Ince, S. D. Kraus, *et al.*, Trophic relationships and oceanography on and around a small offshore bank. *Mar. Ecol. Prog. Ser.*, **363** (2008), 15–28.

Susetiono, *Lamun dan Fauna. Teluk Kuta, Pulau Lombok* (Jakarta: LIPI Press, 2007).

Tittensor, D. P., C. Mora, W. Jetz, *et al.*, Global patterns and predictors of marine biodiversity across taxa *Nature* (2010), doi:10.1038/nature09329.

Trott, T., Cobscook inventory: a historical checklist of marine invertebrates spanning 162 years. *Northeast Nat.*, **11**:Special Issue 2 (2004), 261–324.

Trott, T., Location of biological hotspot in the Gulf of Maine. Gulf of Maine Symposium, 4–9 October 2009, St. Andrews, New Brunswick (in press).

Udvardy, M., *Dynamic Zoogeography* (New York: Van Nostrand Reinhold Company, 1969).

Venkataraman, K. and M. Wafar, Coastal and marine biodiversity of India. *Indian J. Mar. Sci.*, **34**:1 (2005), 57–75.

Vermeij, G. J., G. P. Dietl, and D. G. Reid, The Trans-Atlantic history of diversity and body size in ecological guilds. *Ecology*, **89**:11 (2008), S39–S52.

Veron, J. E. N., *Corals of the World* (Townsville: Australian Institute of Marine Science, 2000).

Watanabe, S., A. Metaxas, J. Sameoto, and P. Lawton, Patterns in abundance and size of two deep-water gorgonian octocorals, in relation to depth and substrate features off Nova Scotia. *Deep Sea Res. I*, **56**:12 (2009), 2235–48.

Watling, L., A review of the genus *Iridogorgia* (Octocorallia: Chrysogorgiidae) and its relatives, chiefly from the North Atlantic Ocean. *J. Mar. Biol. Assoc. UK*, **87**:2 (2007). 393–402.

Watling, L. and C. Skinder, Video analysis of megabenthos assemblages in the Central Gulf of Maine. Special Paper – Geological Association of Canada (2007). 369–77.

Wilkinson, C., *Status of Coral Reefs of the World: 2004* (Townsville, Australia: Australian Institute of Marine Science, 2004).

Willig, M. R., D. M. Kaufman, and R. D. Stevens, Latitudinal gradients of biodiversity: pattern, process, scale, and synthesis. *Annu. Rev. Ecol. Evol. Syst.*, **34** (2003), 273–309.

Witman, J. D., R. J. Etter, and F. Smith, The relationship between regional and local species diversity in marine benthic communities: a global perspective. *Proc. Natl. Acad. Sci. USA*, **101**:44 (2004), 15664–9.

Wlodarska-Kowalczuk, M., P. Kuklinski, M. Ronowicz, J. Legezynska, and S. Gromisz, Assessing species richness of macrofauna associated with macroalgae in Arctic kelp forests (Hornsund, Svalbard). *Polar Biol.*, **32**:6 (2009), 897–905.

Worm, B., M. Sandow, A. Oschlies, H. K. Lotze, and R. A. Myers, Global patterns of predator diversity in the open oceans. *Science*, **309**:5739 (2005), 1365–9.

Yasin, Z., S. Y. Kwang, A. T. Shau-Hwai, and Y. Shirayama, *Field Guide to the Echinoderms (Sea Cucumber and Sea Stars) of Malaysia* (Kyoto, Japan: Kyoto University Press, 2008).

York, A., Habcam sheds light on invading tunicate. *Sea Technol.*, **50**:8 (2009), 41–6.

Zimmerman, T. L. and J. W. Martin, Artificial Reef Matrix Structures (ARMS): an inexpensive and effective method for collecting coral reef-associated invertebrates. *Gulf Caribb. Res.*, **16** (2004), 59–64.

CHAPTER

6

At the ends of the Earth

Although a sphere has no ends, remote Arctic and Antarctic Oceans seem like the ends of the Earth for most of us. Few locations conjure up exploration as the poles do. Freezing in the dark, Amundsen, Peary, Scott, and Shackleton raced to "conquer" the poles first. Only five decades ago Byrd established permanent Antarctic bases. Ice and darkness invoke mystique and challenge. As much as I might grumble, bobbing seasick in the drizzle, fog, and gales off Newfoundland, my discomfort pales next to ArcOD (*Arctic Ocean Diversity*) scientists loading rifles to fend off polar bears that kill with one swipe of a paw. Rifles and every single piece of equipment must be flown or shipped in advance. There is no running to the hardware store for replacement parts, and sailing and flying people and equipment to deserted poles drains budgets. Polar researchers still routinely drag equipment across the ice on sleds just as Norwegian explorer Fridtjof Nansen did a century ago as he collected some of the first data on Arctic ice movement. Snow machines have replaced dogs, but much is the same.

Our glimpses are longer during 24 hour summer sunlight, when protective ice cover retreats and we can see ocean life, but the water and seafloor beneath permanent ice are rarely sampled. New glimpses of spectacular life "through the ice" by Census scientists have been featured in postage stamps, art exhibitions, and worldwide press coverage (Figure 6.1).

Despite few humans reaching the poles, human effects are apparent. Global warming imperils polar marine life, as captured in photographs of polar bears on melting ice floes. Shortly after scientists first described them, hunters extinguished Steller's sea cow (*Hydrodamalis gigas*) and great auk

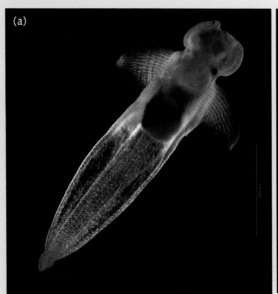

(a)

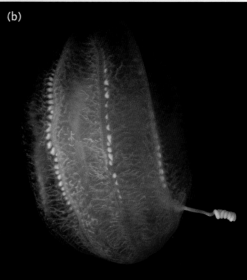

(b)

Figure 6.1 Spectacular polar life near the surface down to the seafloor.

Everywhere they looked at the poles, north and south, scientists found life.

(a) The shell-less pteropod or swimming snail, *Clione limacina*, grows as long as 4 centimeters in both Arctic and Antarctic waters, and in waters in between.

(b) The comb jelly, *Aulacoctena* sp., collected with a remotely operated vehicle (ROV) in the deep Arctic Canada Basin.

(c) An Arctic sea star collected with an ROV on the deep-sea floor of the far northern Canada Basin.

(c)

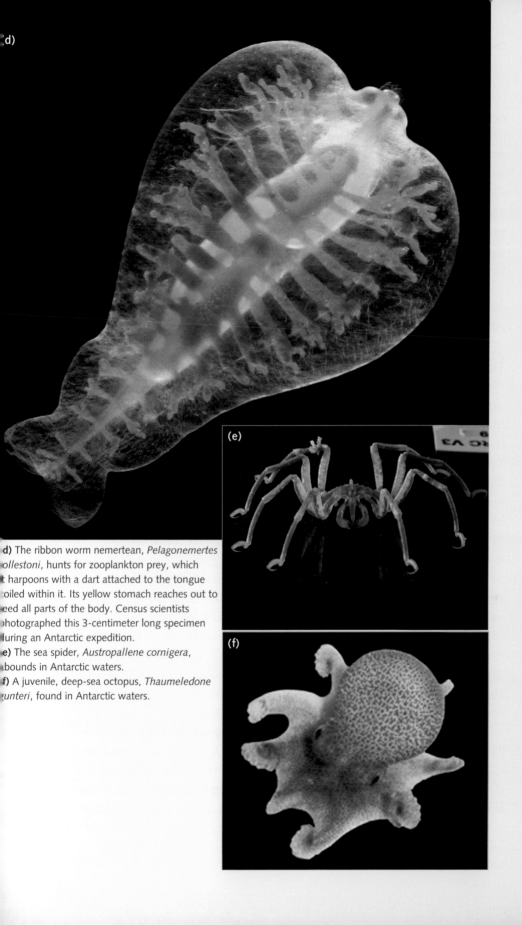

(d)

(e)

(f)

d) The ribbon worm nemertean, *Pelagonemertes rollestoni*, hunts for zooplankton prey, which it harpoons with a dart attached to the tongue coiled within it. Its yellow stomach reaches out to feed all parts of the body. Census scientists photographed this 3-centimeter long specimen during an Antarctic expedition.
e) The sea spider, *Austropallene cornigera*, abounds in Antarctic waters.
f) A juvenile, deep-sea octopus, *Thaumeledone gunteri*, found in Antarctic waters.

(*Pinguinus impennis*), leaving only dusty museum specimens to prove they existed. The Arctic ice cover that hinders scientific access also protects areas from fishers and hunters. If melting opens the Arctic to shipping, exploration, and fishing, marine life will change. Antarctic exploitation hunted fur and elephant seals to near extinction, and diminished global populations of blue, fin, humpback, southern right, and sei whales, some by half and some by 99%. The collapse of fisheries in accessible seas in the 1960s drove bottom trawlers to more remote areas. As bottom habitat was simultaneously destroyed, stocks of Antarctic marbled rockcod, *Notothenia rossii*, and mackerel icefish, *Champsocephalus gunnari*, quickly declined.

A tale of two oceans

The 19 Arctic ecoregions and 21 Antarctic ecoregions comprise just 17% of the total 232 global ecoregions. Nevertheless, the ice and darkness overwhelming their annual cycles singles them out, encompassing habitats from the intertidal to the deep sea, and pelagic environments in between (Figure 6.2). Land surrounds the ice-covered Arctic Ocean, whereas the Southern Ocean surrounds the ice-covered Antarctic land (Figure 6.2 inset). Both have summers of continuous light, winters of continuous darkness, and major habitat defined by ice. Scattered stations of various nations created hot spots of ecological knowledge, some going back hundreds of years. In the Antarctic, different nations run 45 seasonal or year-round stations leaving wide gaps of knowledge in between.

The knowledge gaps made Arctic and Antarctic research projects an absolute must for the Census, and in 2007 the International Polar Year focused public interest and opened opportunities for collaboration. Because the ArcOD and *Census of Antarctic Marine Life* (CAML) projects spanned remote habitats, they encouraged collaboration within and outside the Census. Although not rich in species, polar life is attuned to cold, seasonal light, and ice, and their cascading effects. Distinct species have evolved to live in Arctic and Antarctic waters, but some live at both poles.

The Arctic Ocean occupies only 4% of the global ocean, smaller than the 23% of the Atlantic and 46% of the Pacific Oceans. At 6% the Southern Ocean is just slightly larger. The land surrounding the Arctic Ocean creates some of the widest continental shelf in the world. More than half of the Arctic Ocean is less than 200 meters deep, and the modest Arctic Ocean holds 31% of all

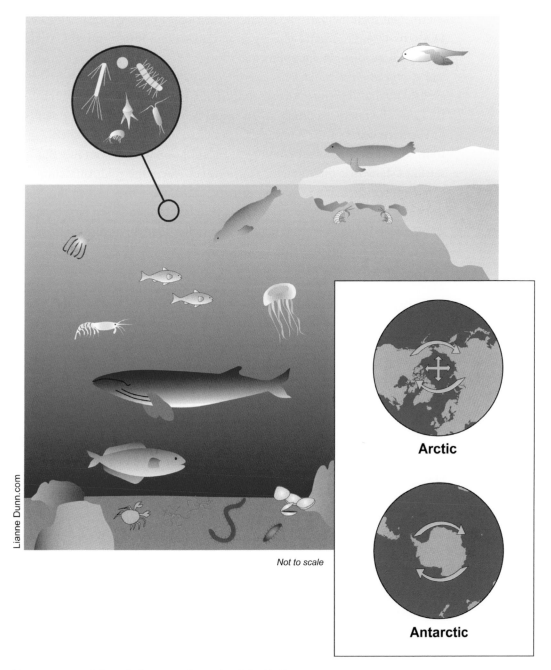

Not to scale

Arctic

Antarctic

Figure 6.2 Ice defines habitats at the ends of the Earth.
This drawing shows the special characteristics of polar ecosystems. Near the
poles, ice reaching out from coasts defines seasons and thus defines changes in
food and habitat. The inset illustrates the differences between the Arctic Ocean,
surrounded by land, and the Southern Ocean, surrounding Antarctic land.
Arrows illustrate potential exchange.

continental shelf habitat. The Antarctic shelf raises the total polar shelf by 11.4% to 42%. Freshwater flowing from Canada and Russia into the Arctic Ocean creates distinct environments, strongly affected by season, circulation, and *stratification* of water. This layering of fresh and salt water limits nutrient availability and therefore Arctic productivity. Freshwater melting off the land into the Southern Ocean is less important to ocean life.

The deep water of the Southern Ocean separates the Antarctic shelf from shelves around other continents, isolating Antarctic Ocean life. But satellites show that the Antarctic Convergence where the cold Antarctic Circumpolar Current and warm northern waters meet may breach the isolation. Circulation gyres move algae, zooplankton, and even larval stages across this boundary, leaving temperature rather than any hydrographic barrier to isolate Antarctic life.

The polar past

When the Bering Strait opened up about five million years ago, Pacific species moved in and mixed with Arctic species. The narrow Bering Strait between Alaska and Russia connects the Arctic Ocean to the Pacific and the Fram Strait between Greenland and Norway's Spitsbergen Island connects the Atlantic to the Arctic. Species from the Atlantic cross the Fram Strait, adding to a mix of deep-sea cosmopolitan species and endemic species that evolved in the Arctic. Over eons, circulation, geology, and glaciation opened and closed connections, and endemic species evolved when connections closed. ArcOD scientists are assembling a biogeographic history of the Arctic that links these events to today's biota and helps predict tomorrow's ocean.

Molecular tools help reveal evolutionary history. The "*molecular clock*" concept dating to the early 1960s assumes molecular change is proportional to time and dates evolutionary events from evidence of genetic sequences. The molecular clock helps differentiate migrants from species that evolved separately in the polar oceans. Different lineages evolved between the shallow Antarctic and sub-Antarctic regions about five million years ago, from the urchin *Sterechinus*, to the brittle star *Astrotoma agassizii*, to the bivalve *Limatula*, to the limpet *Nacella*. Over eons, species evolved in the Antarctic and spread northward into deep water, through *tropical submergence*. The deep Southern Ocean may have pumped out diverse crustaceans and anemones. Thirty million years ago, some deep-sea octopods evolved from a common

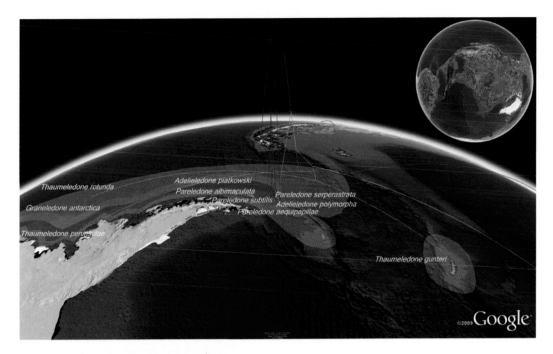

Figure 6.3 The Antarctic species incubator.
Visualization of how different species of octopus evolved from Antarctic ancestors and moved northward, some farther than others. The red line on the inset shows that some species of Antarctic octopus crossed the tropics and reached far into the northern hemisphere, all the way to Greenland and Alaskan waters.

Antarctic ancestor and moved north (Figure 6.3), as have isopod crustaceans. The Antarctic amphipod crustacean genus *Liljeborgia* has eyes, but related species that have moved into the lightless deep sea are blind. During long evolutionary periods, some isopods without eyes "emerged" from the deep sea into shallow water, through *polar emergence*. The development of 29 lineages of sea slug, *Austrodoris kerguelenensis* with Antarctic ancestry hints that the "discontinuity" of the Antarctic Convergence may be less critical than once thought.

Fauna in shallow Antarctic water differs from that on the adjacent slope and abyssal plain, and across the Antarctic Convergence. How did this diverse fauna evolve? Many Antarctic species evolved when glaciers interrupted the continental shelf and grounded ice broke the shelf into isolated "bays," where species evolved separately by *allopatric* speciation. But the present shelf is more or less continuous and species intermingle so that diverse octopus, isopod crustaceans, and sea cucumbers coexist.

The icy waters of polar present

In the polar present, ice that covers the entire Arctic Ocean during winter retreats by half during summer. As winter's perpetual darkness gives way to summer's perpetual light, temperatures warm and diatoms dominate algae in the ice, contributing as much as half the primary productivity in the central Arctic. Ice and snow still block out all but a few percent of any light that might penetrate to water beneath. Nevertheless, the little light passing through during springtime melting energizes the *sympagic* community living at the ice–water interface (Figure 6.4) and within salty brine channels that permeate the ice. Feast follows winter famine as the algae multiply and invertebrates colonize and feed in this unique world where water in brine channels may be three times saltier than seawater and thus remain fluid at −10 °C without freezing.

Although the extreme conditions in ice are too much for most life, those species that can cope enjoy an "all you can eat" buffet of algae on the underside of the ice using the available dim light (Figure 6.4). Low diversity, but high abundances, of protists and meiofauna, as well as larval stages of benthic species, live and feed on algae in the rich matrix. Different species, some endemic, can live in the dynamic, three-dimensional ice, including newly discovered predators. Fecal pellets from the feeding frenzy sink and feed other species. Arctic cod eat amphipods and other ice invertebrates, and they, in turn, feed seals and other species in the food web. Buckling ice heaves up ridges that extend more than 20 meters below, adding complexity. CAML also found important effects of ice melt on Antarctic plankton, though they studied ice melt dynamics less than ArcOD.

In early Arctic July, warming melts and breaks up ice, thus releasing algae. The movable feast then transfers to the water column. As sea ice and algae fluctuate, so does faunal abundance and diversity. By summer's peak, the endless days deliver more light to Arctic waters than reaches sunny Caribbean beaches. But light still limits because the angle of the sun is low, and soon nutrients become depleted and so productivity is brief.

While food is abundant, the response is rapid. Just as bears on land store fat for winter hibernation, the zooplankton in the sea store "food" in lipid form, then respond quickly when new food arrives with the spring bloom. Growth rates in zooplankton are strongly seasonal, and many species eat whatever they can get. Both growth and depth distribution of zooplankton fluctuate seasonally, altering energy transfer between surface and deeper layers. The daily vertical zooplankton commutes common in many different ecosystems (see Chapter 7) stop during 24-hour Arctic sunlight.

In the feeding season, abundant seabirds, seals, and whales such as bowhead, gray, and beluga congregate at oceanographic fronts where water masses meet. They feast in numbers that rival those anywhere on Earth. In the Arctic the feast spans a few dominant copepod species, such as *Calanus finmarchicus*, to capelin, *Mallotus villosus*, and herring, *Clupea harengus*, and on to whales, seals, and seabirds. Euphausiids, or krill, fill a key part of this role in the Antarctic.

Scientists know the dominant phytoplankton and zooplankton that fuel the Arctic biota. Much species data, however, were scattered in file cabinets in Russia, the United States, Canada, and Europe. When ArcOD convened five years ago, only some 300 species of zooplankton and 300 species of phytoplankton were known. ArcOD planned how to fill gaps by focusing on poorly known regions and groups of organisms, which now fill published papers and new taxonomic guides in development.

In summer, glaciers cover only 20% of the shelf around the central Antarctic land mass, but in winter extend to 60%, doubling the size of the Antarctic continent and extending beyond the continental shelf over the deep ocean. An ample nutrient supply does not make a productive Southern Ocean because iron limits phytoplankton production, except where islands naturally inject iron. Commercial companies have proposed *iron fertilization*, adding iron to bloom phytoplankton, in the Southern Ocean so they take up more atmospheric carbon dioxide.

Scientists know the simple links of phytoplankton to zooplankton to fish, whales, seals, seabirds, and seafloor detritus feeders. But they also know that nutrient and energy cycling through microbes in the *microbial loop* helps sustain the food web in the absence of abundant phytoplankton.

The convergence between the Antarctic Circumpolar Current and northerly waters divides pelagic species between north and south. CAML refined this pattern by finding varied plankton and fishes that differ in diversity and abundance within the Circumpolar Current. Smaller zooplankton have begun to replace the euphausiid Antarctic krill (*Euphausia superba*) that is a widespread, important crustacean in the Antarctic pelagic food web, but has declined significantly. CAML also discovered that Antarctic fishes called *myctophids*, or lanternfish, transfer energy from 100 meters to 1,000 meters in their daily commutes.

Figure 6.4 Habitats at the ends of the Earth.
Though not especially diverse, the span from ice-covered surface down to the deep seafloor encompasses an array of polar habitats that support their own suites of species.
(a) On the ice in the Arctic Canada Basin, a polar bear mother and her cub.
(b) Below the Arctic ice, ice amphipods feed on algae attached to ice crystals.
(c) In the Antarctic Larsen B region, deep-water corals grow in abundance at 120 meters depth.

(d)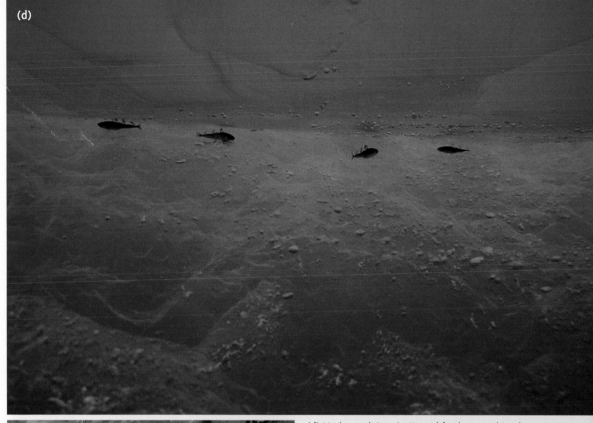

(d) Under pack ice, Arctic cod feed on amphipods and other ice fauna.

(e) In the Antarctic, icefish adapted to cold have no hemoglobin or red blood cells, which makes their blood thinner so it pumps more easily and saves energy.

(f) Along Alaska's Aleutian Archipelago, *Ptychodactis* sp., a new species of "swimming" anemone. They detach from the hard substrate, inflate, and drift with currents over the seafloor to attach elsewhere.

The polar present seafloor

Sand and gravel cover the inner Arctic shelves while mud covers the deep seafloor. After ice retreats, the feast moves to the seafloor because slow metabolism in cold water limits how quickly zooplankton can respond to the rapid spring bloom. The high biomass on some shelf seafloor, dominated by dense brittle stars, amphipod crustaceans, polychaete worms, and mollusks, fades downward so that deep-sea communities are numerically sparse, dominated by *deposit feeders* that strip food material from ingested sediment particles. The scarce, but fresh, food that reaches the seafloor links seafloor abundance to planktonic production. Because 90% of known species live on the seafloor, it is something of a hot spot for Arctic diversity.

The Antarctic continental shelf averages 450 meters deep and sometimes extends out beyond 1,000 meters. As in the Arctic, glacial rock, gravel, and mud formed from silica shells of sinking diatoms cover the seafloor (Figure 6.4). A rich diversity of bryozoans, sponges, and amphipods lives on the shelf, but some mollusks, isopod and decapod crustaceans, and fishes are absent despite presence in fossils. Some groups of Antarctic mollusks and crustaceans are well known, whereas nematodes are not. Although past estimates of 80% probably inflate the level of Antarctic endemism, new reports of half or more in some groups are probably more realistic.

Along the Antarctic shelf, macrofauna encompass one large ecoregion, some stretches are productive and rich in species, where others are sparse. Areas previously, or recently, scoured by icebergs or perpetually beneath ice typically have low abundances and diversity because of limited food.

A plethora of new ocean life

The 60 (and counting) new Arctic species that ArcOD has discovered span from microscopic to *megafaunal* animals of a size seen easily in bottom photos, in habitats from ice channels through the water column to the seafloor (Figure 6.4). They found new species in little-known regions such as the Canada Basin, and in novel habitats such as the area of reduced flow above the seafloor called the *benthic boundary layer*. Discoveries spanned a predatory hydrozoan from brine channels, poorly studied jellyfish, and yet to be described copepod, amphipod, and ostracod crustaceans from the water column. CAML sampling increased the number of Antarctic gelatinous zooplankton species by a factor of two to three times. ArcOD also found viruses thriving in all sorts of Arctic environments.

On the seafloor, ArcOD found new species of snails, polychaete worms, copepod crustaceans, and sea cucumbers. They found new species of *bryozoans*, which are coral-like animals with calcium carbonate skeletons. Many other new species were discovered and await formal naming. The Chukchi Sea added over 300 species to a 2001 inventory, including 21 polychaete worm species previously unknown in the Arctic.

The collaboration between the Census CAML and (*Census of Diversity of Abyssal Marine Life*) CeDAMar projects collected 1,400 species in the deep Antarctic including perhaps 700 that are new. A group of protist *komokiaceans*, a type of foraminifera first formally described in 1977, were unknown from the region until the Antarctic cruises collected 50 species, 35 new. The 674 or 87% new isopod crustaceans among Census collections added 15% to the global total for that group. The Census also discovered new species of glass sponges, carnivorous sponges, and calcareous sponges. They found many *epibionts*, where one species provides habitat for others and thus adds diversity. For example, the spines of one sea urchin housed 156 different species and the blades of Arctic kelp housed a mixture of attached epibionts.

Exploration and environmental change have expanded the known ranges of even familiar groups such as fishes. Harbingers of the consequence of warming, walleye pollock, *Theragra chalcogramma*, were found 200 kilometers farther north in the Arctic and surveys sighted rare toothed whales closer to Antarctica than ever seen. Workers in the Canada Basin provided new data on marine mammals and birds. Others resolved taxonomy problems in coral-like bryozoans and jellyfish. Not surprisingly, extensive sampling collected the largest numbers.

Pooling polar knowledge

Antarctic programs that had brought researchers together expedited CAML's 2005 launch. The *Scientific Committee on Antarctic Research*, SCAR, has coordinated international Antarctic research for almost 50 years, and helped CAML hit the ground running. They coordinated research efforts and created a biodiversity database (SCAR-MarBIN) linked to OBIS.

From 2007 to 2009, 34 nations and 321 scientists on 18 vessels sailed on International Polar Year and CAML cruises, an impressive effort in vessels that cost many tens of thousands of US dollars daily to operate. As of May 2009, the database contained over one million records of 8,000 species from 122 datasets. This doubles the numbers from 1993.

As of August 2009, ArcOD had delivered 26 Arctic datasets to OBIS, including 150,000 data records and was expected to surpass 250,000 by the end of 2010. ArcOD contributed to a field manual for sampling ice environments, and has worked with Russia's Zoological Institute in St. Petersburg to initiate a set of taxonomic guides for Arctic marine life that is just now beginning to appear.

Using barcoding technologies adapted from ICoMM, ArcOD and CAML discovered 1,500 kinds of Arctic bacteria and 700 kinds of archaea. They barcoded 300 species of Arctic benthos, 93 species of fishes, and 41 species of Arctic zooplankton in collaboration with CMarZ. CAML has added over 11,000 Antarctic barcode sequences from their collections. In one study CAML examined over 25,000 plankton barcodes, where 13,000 were new sequences.

Over a century ago Captain James Clark Ross proposed *bipolar species* that live in both poles. Some species, such as the sooty shearwater, live at both poles because they migrate between. Searching their databases ArcOD and CAML found 230 species names common to both poles. Truly bipolar species don't include cosmopolitan species that occur elsewhere, often in cold, deep water, or species that look the same and share a name, but are genetically distinct.

An international team identified 227 species in 12 phyla as potentially bipolar. Other historical records suggested another 102. A strategy of assembling known distributions and a strict definition of polar eliminated about 80% of these species. Most of the remaining 20% were cosmopolitan with only a few not found in between. Thus, truly bipolar species are rare, encompassing only 5% of a list of the most promising candidates. Although these few are currently considered bipolar, some may eventually be found in the poorly sampled deep sea. Genetic analysis mostly found genetic differences and cryptic forms rather than bipolar species. Thus, bipolar species that live at the opposite poles, but nowhere else, are few.

The Arctic is considered species poorer than the Antarctic for geological reasons, and bacterial comparisons support this view. A new survey of biodiversity knowledge for Canada's Atlantic, Pacific, and Arctic waters found Arctic microbial diversity highest, but this pattern may reflect more intense Arctic sampling. Warmer water and habitat diversity enhance Antarctic biodiversity. Nevertheless, the current total of about 8,200 Antarctic versus 6,000 Arctic multicellular species, or *metazoans*, shows a modest difference between the poles.

The polar future

What's left to find? An exploration of Arctic marine biodiversity led by the European Union found over 1,400 species from a 50-square-kilometer area less than 280 meters deep. Species accumulation curves show that mollusks are almost fully described, but other groups – particularly smaller-sized organisms – are poorly known. ArcOD estimates an eventual discovery of 45,000 kinds of bacteria, 5,000 kinds of archaea, and 4,500 species of protists. Arctic microbes, once thought inactive, may not be and archaea, bacteria, and protists thrive in all types of Arctic habitat.

The Antarctic intertidal, once thought species poor, may instead be diverse and productive. The shelf seafloor alone may have 17,000 macrofaunal species, and that doesn't even include the spectacular deep-ocean diversity sampled by CeDAMar. The 90% of the region deeper than 1,000 meters also holds many unknowns!

Explorers know little about the deep basins that extend thousands of kilometers across the Arctic seafloor. What they do know shows key differences in species composition and abundance between sides and tops of ridges. The newly discovered abundant hydrothermal vent sites on the Gakkel Ridge and geologically "recent" and explosive volcanic activity illustrate the new environments and species yet to be discovered in remote parts of the Arctic. Unexplored areas of the Antarctic continental shelf covered by floating ice hold other new discoveries, as do the Amundsen and Western Weddell Sea.

The human stain

Warming temperatures and thinning ice add to concern about Arctic summer multi-year ice that is declining by 8.6% per decade. These changes might drive endemic Arctic species extinct, while others might escape northward from warmer waters farther south, perhaps actually increasing Arctic diversity, even as ice-dependent species go extinct. The few long-term studies of Arctic intertidal fauna, zooplankton, and benthos indicate a constant species pool, but changing abundance of dominant species. With the arrival of new species and the warming of waters, sorting out Arctic change will be difficult. Continued sampling will be critical. Pacific seafloor species have recently appeared in the Chuckchi, perhaps in climate-linked invasions. If seabirds that feed on plankton and fish starve as food disappears, oceanic changes will affect the land.

Some of the most rapid warming on Earth has occurred in areas of the Antarctic. In recent decades, surface waters have warmed by 1°C, sea-ice formation is 10% less, and ice shelves have collapsed, causing major changes on the Antarctic Peninsula. Ocean acidification is expected to hit the Antarctic first, because a wide range of organisms that use calcium carbonate may be seriously affected by the year 2100.

Arctic exploration and oil development are rapidly expanding. The mercury concentrations in whales that reflect their feeding is already an environmental and health issue. If predictions of an ice-free Arctic in the summer months by 2030 or 2100 come true, shipping traffic, oil exploration, and fishing activity will escalate. To understand how these activities might affect Arctic marine life, ArcOD advocated the expansion of several time-series sampling programs in sensitive regions in the Chukchi and Barents Sea where change is happening quickly. New Arctic observatories could monitor changes in real time and from afar.

After whaling ships plied the water and pushed marine mammals and seabirds to near extinction, their impacts on the Antarctic have decreased. Fisheries for krill, toothfishes, and icefish continue to land hundreds of thousands of tons, and krill removal has caused concern. But a central regulatory body for Antarctic fisheries is working to ensure that food webs are not significantly altered by these removals, though illegal fisheries are difficult to control in such a remote area. The double-edged sword of tourism has, on the one hand, heightened public awareness of the Antarctic ecosystem, but on the other hand increased from 7,000 visitors in 1992 to 35,000 in 2007, raising concerns about possible disturbances.

CAML argues that "Antarctic waters might be the best protected marine areas on Earth" in large part because they are shared by multiple nations and despite their remoteness benefit from the public view of their pristine condition. Nations claiming parts of the Antarctic for their economic potential focus research on their claim areas. South American and European nations focus on the West Antarctic and the Scotia and Weddell Seas, whereas Australian, New Zealand, and Asian nations focus on the East Antarctic. Russia and the United States operate in both areas. Bioprospecting, climate-change effects on ecosystem functions, and biodiversity loss point to a need for long-term observations similar to those conducted by the Census. Collectively they could become a Southern Ocean Observation System to signal changes in sensitive areas like the Antarctic Convergence and the Antarctic deep sea.

BIBLIOGRAPHY

Allcock, A. L., On the confusion surrounding *Pareledone charcoti* (Joubin, 1905) (Cephalopoda: Octopodidae): endemic radiation in the Southern Ocean. *Zool. J. Linn. Soc.*, **143**:1 (2005), 75–108.

Allcock, A. L., R. R. Hopcroft, J. Strugnell, *et al.*, Biopolarity in marine invertebrates: myth or marvel (unpublished manuscript).

Allcock, A. L., J. M. Strugnell, P. Prodohl, U. Piatkowski, and M. Vecchione, A new species of *Pareledone* (Cephalopoda: Octopodidae) from Antarctic Peninsula waters. *Polar Biol.*, **30**:7 (2007), 883–93.

Archambault, P., P. V. R. Snelgrove, J. A. D. Fisher, *et al.*, From sea to sea: Canada's three oceans of biodiversity. *PLoS ONE*, doi:10.1371/journal.pone.0012182.

Arntz, W. E., T. Brey, and V. A. Gallardo, Antarctic zoobenthos. *Oceanogr. Mar. Biol. Annu. Rev.*, **32** (1994), 241–304.

Atkinson, A., V. Siegel, E. A. Pakhomov, *et al.*, Oceanic circumpolar habitats of Antarctic krill. *Mar. Ecol. Prog. Ser.*, **362** (2008), 1–23.

Atkinson, A. V. Siegel, E. Pakhomov, and P. Rothery, Long-term decline in krill stock and increase in salps within the Southern Ocean. *Nature*, **432**:7013 (2004), 100–3.

Azam, F., D. C. Smith, and J. T. Hollibaugh, The role of the microbial loop in Antarctic pelagic ecosystems. *Polar Res.*, **10** (1991), 239–44.

Barnes, D. K. A., D. A. Hodgson, P. Convey, C. S. Allen, and A. Clarke, Incursion and excursion of Antarctic biota: past, present and future. *Glob. Ecol. Biogeogr.*, **15**:2 (2006), 121–42.

Berning, B. and P. Kuklinski, North-east Atlantic and Mediterranean species of the genus *Buffonellaria* (Bryozoa, Cheilostomata): implications for biodiversity and biogeography. *Zool. J. Linn. Soc.*, **152**:3 (2008), 537–66.

Beuchel, F., B. Gulliksen, and M. L. Carroll, Long-term patterns of rocky bottom macrobenthic community structure in an Arctic fjord (Kongsfjorden, Svalbard) in relation to climate variability (1980–2003). *J. Mar. Syst.*, **63**:1–2 (2006), 35–48.

Blachowiak-Samolyk, K., S. Kwasniewski, K. Dmoch, H. Hop, and S. Falk-Petersen, Trophic structure of zooplankton in the Fram Strait in spring and autumn 2003. *Deep Sea Res. II*, **54**:23–26 (2007), 2716–28.

Blachowiak-Samolyk, K., S. Kwasniewski, K. Richardson, *et al.*, Arctic zooplankton do not perform diel vertical migration (DVM) during periods of midnight sun. *Mar. Ecol. Prog. Ser.*, **308** (2006), 101–16.

Bluhm, B. A., K. O. Coyle, B. Konar, and R. Highsmith, High gray whale relative abundances associated with an oceanographic front in the south-central Chukchi Sea. *Deep Sea Res. II*, **54**:23–26 (2007), 2919–33.

Bluhm, B. A. and R. Gradinger, Regional variability in food availability for Arctic marine mammals. *Ecol. Appl.*, **18**:2 (2008), S77–S96.

Bluhm, B. A., R. Gradinger, and S. Piraino, First record of sympagic hydroids (Hydrozoa, Cnidaria) in Arctic coastal fast ice. *Polar Biol*, **30**:12 (2007), 1557–63.

Bluhm, B. A., R. Gradinger, and S. Schnack-Schiel, Sea ice meio- and macrofauna. In D. Thomas and G. Dieckmann (eds.), *Sea Ice: An Introduction to its Physics, Chemistry, Biology and Geology* (New York: Wiley-Blackwell, 2009), pp. 357–93.

Bluhm, B. A., K. Iken, and R. R. Hopcroft, Observations and exploration of the Arctic's Canada Basin and the Chukchi Sea: the hidden ocean and RUSALCA expeditions. *Deep Sea Res. II*, **57** (2010), 1–4.

Booth, B. C. and R. A. Horner, Microalgae on the Arctic Ocean Section, 1994: species abundance and biomass. *Deep Sea Res. II*, **44**:8 (1997), 1607–22.

Brandt, A., C. De Broyer, I. De Mesel, *et al.*, The biodiversity of the deep Southern Ocean benthos. *Philos. Trans. R. Soc. Lond. B Biol. Sci.*, **362**:1477 (2007), 39–66.

Brandt, A., A. J. Gooday, S. N. Brandao, *et al.*, First insights into the biodiversity and biogeography of the Southern Ocean deep sea. *Nature*, **447**:7142 (2007), 307–11.

Bucklin, A. and B. W. Frost, Morphological and molecular phylogenetic analysis of evolutionary lineages within *Clausocalanus* (Copepoda: Calanoida). *J. Crustacean Biol.*, **29**:1 (2009), 111–20.

Carlsen, B. P., G. Johnsen, J. Berge, and P. Kuklinski, Biodiversity patterns of macro-epifauna on different lamina parts of *Laminaria digitata* and *Saccharina latissima* collected during spring and summer 2004 in Kongsfjorden, Svalbard. *Polar Biol.*, **30**:7 (2007), 939–43.

Chaban, E. M., *Opistobranchiate mollusca* of the orders Cephalaspidea, Thecosomoata and Gymnosomata (Mollusca, Ophisthobranchia) of the Chukchi Sea and Bering Strait. Fauna and zoography of benthos of the Chukchi Sea. *Explor. Fauna Seas*, **61**:69 (2008), 149–62.

Cheung, W. W. L., V. W. Y. Lam, J. L. Sarmiento, *et al.*, Projecting global marine biodiversity impacts under climate change scenarios. *Fish Fish.*, **10**:3 (2009), 235–51.

Clarke, A., Temperature and evolution: Southern Ocean cooling and the Antarctic marine fauna. In K. R. Kerry and G. Hempel (eds.), *Antarctic Ecosystems: Ecological Change and Conservation* (New York: Springer, 1990), pp. 9–22.

Clarke, A., D. K. A. Barnes, and D. A. Hodgson, How isolated is Antarctica? *Trends Ecol. Evol.*, **20**:1 (2005), 1–3.

Clarke, A. and J. A. Crame, The origin of the Southern Ocean marine fauna. In J. A. Crame (ed.), *Origins and Evolution of the Antarctic Biota* (London: The Geological Society, 1989), pp. 253–68.

Clarke, A. and N. M. Johnston, Antarctic marine benthic diversity. *Oceanogr. Mar. Biol. Annu. Rev.*, **41** (2003), 47–114.

Connelly, T. L., Biogeochemistry of benthic boundary layer zooplankton and particulate organic matter on the Beaufort Sea shelf. PhD thesis, Memorial University of Newfoundland (2008).

Cristobo, F. J., V. Urgorri, and P. Rios, Three new species of carnivorous deep-sea sponges from the DIVA-1 expedition in the Angola Basin (South Atlantic). *Org. Divers. Evol.*, **5**:Suppl. 1 (2005), 203–13.

Dayton, P. K., B. J. Mordida, and F. Bacon, Polar marine communities. *Am. Zool.*, **34** (1994), 90–9.

De Broyer, C. and B. Danis, How many species in the Southern Ocean? Towards a dynamic inventory of Antarctic marine species. *Deep Sea Res. II* (in press).

Diaz, A., J. P. Feral, B. David, T. Saucede, and E. Poulin, Evolutionary pathways among shallow and deep sea echinoids of the genus *Sterechinus* in the Southern Ocean. *Deep Sea Res. II* (in press).

Edmonds, H. N., P. J. Michael, E. T. Baker, *et al.*, Discovery of abundant hydrothermal venting on the ultraslow-spreading Gakkel Ridge in the Arctic. *Nature*, **421**:6920 (2003), 252–6.

Eicken, H., R. Gradinger, A. Gaylord, *et al.*, Sediment transport by sea ice in the Chukchi and Beaufort Seas: increasing importance due to changing ice conditions? *Deep Sea Res. II*, **52**:24–26 (2005), 3281–302.

Eicken, H., A. L. Lovecraft, and M. L. Druckenmiller, Sea-ice system services: a framework to help identify and meet information needs relevant for Arctic observing networks. *Arctic*, **62**:2 (2009), 119–36.

Fuhrman, J. A., J. A. Steele, I. Hewson, *et al.*, A latitudinal diversity gradient in planktonic marine bacteria. *Proc. Natl. Acad. Sci. USA*, **105**:22 (2008), 7774–8.

Gagaev, S., *Terebellides irinae* sp. n. – a new species of the genus *Terebellides* (Polychaeta, Terebellidae) from the Arctic Basin. *Russ. J. Mar. Biol.*, **35** (2009), 474–8.

Gagaev, S. Y., *Sigambra healyae sp n.*, a new species of polychaete (Polychaeta: Pilargidae) from the Canadian Basin of the Arctic Ocean. *Russ. J. Mar. Biol.*, **34**:1 (2008), 73–5.

Galand, P. E., E. O. Casamayor, D. L. Kirchman, M. Potvin, and C. Lovejoy, Unique archaeal assemblages in the Arctic Ocean unveiled by massively parallel tag sequencing. *ISME J.*, **3**:7 (2009), 860–9.

Gonzalez Wevar, D., B. David, and E. Poulin, Phylogeography and demographic inference in *Nacella* (*Patinigera*) *concinna* (Strebel, 1908) in the western Antarctic Peninsula. *Deep Sea Res. II* (in press).

Gooday, A. J., T. Cedhagen, O. E. Kamenskaya, and N. Cornelius, The biodiversity and biogeography of komokiaceans and other enigmatic foraminiferan-like protists in the deep Southern Ocean. *Deep Sea Res. II*, **54**:16–17 (2007), 1691–719.

Gosselin, M., M. Levasseur, P. A. Wheeler, R. A. Horner, and B. C. Booth, New measurements of phytoplankton and ice algal production in the Arctic Ocean. *Deep Sea Res. II*, **44**:8 (1997), 1623.

Gradinger, R., Sea ice microorganisms. In G. E. Bitten (ed.), *Encyclopedia of Environmental Microbiology* (New York: Wiley & Sons, Inc., 2002), pp. 2833–44.

Gradinger, R. R. and M. E. M. Baumann, Distribution of phytoplankton communities in relation to the large-scale hydrographical regime in the Fram Strait. *Mar. Biol.*, **111**:2 (1991), 311–21.

Gradinger, R. and B. Bluhm, Arctic ocean exploration 2002. *Polar Biol.*, **28**:3 (2005), 169–70.

Gradinger, R. R. and B. A. Bluhm, In-situ observations on the distribution and behavior of amphipods and Arctic cod (*Boreogadus saida*) under the sea ice of the High Arctic Canada Basin. *Polar Biol.*, **27**:10 (2004), 595–603.

Gradinger, R., B. A. Bluhm, R. R. Hopcroft, *et al.*, Marine life in the Arctic. In A. D. McIntyre (ed.), *Life in the World's Oceans: Diversity, Distribution, and Abundance* (Oxford: Blackwell Publishing Ltd., 2010, pp. 183–202.

Grant, R. A. and K. Linse, Barcoding Antarctic biodiversity: current status and the CAML initiative, a case study of marine invertebrates. *Polar Biol.*, **32**:11 (2009), 1629–37.

Grebmeier, J. M. and J. P. Barry, The influence of oceanographic processes on pelagic-benthic coupling in polar regions: a benthic perspective. *J. Mar. Syst.*, **2**:3–4 (1991), 495–518.

Grebmeier, J. M., L. W. Cooper, H. M. Feder, and B. I. Sirenko, Ecosystem dynamics of the Pacific-influenced Northern Bering and Chukchi Seas in the Amerasian Arctic. *Prog. Oceanogr.*, **71**:2–4 (2006), 331–61.

Griffiths, H. J., Antarctic marine biodiversity – what do we know about the distribution of life in the Southern Ocean? *PLoS ONE* **5**:8 (2010), e11683, doi : 10.1371/journal.pone.0011683.

Griffiths, H. J., D. K. A. Barnes, and K. Linse, Towards a generalized biogeography of the Southern Ocean benthos. *J. Biogeogr.*, **36**:1 (2009), 162–77.

Griffiths, H. J., B. Danis, and A. Clarke, Quantifying Antarctic marine biodiversity: the SCAR-MarBIN data portal. *Deep Sea Res. II* (2010).

Gutt, J., Antarctic macro-zoobenthic communities: a review and an ecological classification. *Antarct. Sci.*, **19**:2 (2007), 165–82.

Gutt, J., I. Barratt, E. Domack, *et al.*, Biodiversity change after climate-induced ice-shelf collapse in the Antarctic. *Deep Sea Res. II* (in press).

Gutt, J., G. Hosie, and M. Stoddart, Marine life in the Antarctic. In A. D. McIntyre (ed.), *Life in the World's Oceans: Diversity, Distribution, and Abundance* (Oxford: Blackwell Publishing Ltd., 2010), pp. 203–20.

Gutt, J. and D. Piepenburg, Scale-dependent impact on diversity of Antarctic benthos caused by grounding of icebergs. *Mar. Ecol. Prog. Ser.*, **253** (2003), 77–83.

Gutt, J., B. Sirenko, I. Smirnov, and W. Arntz, How many macrobenthic species might inhabit the Antarctic shelf? *Antarct. Sci.*, **16** (2004), 11–6.

Harwood, L. A., F. McLaughlin, R. M. Allen, J. Illasiak, and J. Alikamik, First-ever marine mammal and bird observations in the deep Canada Basin and Beaufort/Chukchi seas: expeditions during 2002. *Polar Biol.*, **28**:3 (2005), 250–3.

Held, C., Molecular evidence for cryptic speciation within the widespread Antarctic crustacean *Ceratoserolis trilobitoides* (Crustacea, Isopoda). In A. H. Huiskes, W. W. Gieskes, J. Rozema, R. M. L. Schorno, S. M. van der Vies, and W. J. Wolff (eds.), *Antarctic Biology in a Global Context* (Leiden: Backhuys Publishers, 2003), pp. 135–9.

Held, C. and J. W. Wagele, Cryptic speciation in the giant Antarctic isopod *Glyptonotus antarcticus* (Isopoda: Valvifera: Chaetiliidae). *Sci. Mar.*, **69** (2005), 175–81.

Hempel, G., Antarctic marine food webs. In W. R. Seigfried, P. R. Condy, and R. M. Laws (eds.), *Antarctic Nutrient Cycles and Food Webs* (Heidelberg: Springer-Verlag, 1985), pp. 266–70.

Hop, H., S. Falk-Petersen, H. Svendsen, *et al.*, Physical and biological characteristics of the pelagic system across Fram Strait to Kongsfjorden. *Prog. Oceanogr.*, **71**:2–4 (2006), 182–231.

Hopcroft, R. R., K. N. Kosobokova, and A. I. Pinchuk, Zooplankton community patterns in the Chukchi Sea during summer 2004. *Deep Sea Res. II*, **57** (2010), 27–39.

Horner, R., *Sea Ice Biota* (Boca Raton, FL: CRC Press, 1985).

Hunt, B. P. V. and G. W. Hosie, The Continuous Plankton Recorder in the Southern Ocean: a comparative analysis of zooplankton communities sampled by the CPR and vertical net hauls along 140° E. *J. Plankton Res.*, **25**:12 (2003), 1561–79.

Hunt, B. P. V. and G. W. Hosie, Continuous Plankton Recorder flow rates revisited: clogging, ship speed and flow meter design. *J. Plankton Res.*, **28**:9 (2006), 847–55.

Hunter, R. L. and K. M. Halanych, Evaluating connectivity in the brooding brittle star *Astrotoma agassizii* across the Drake Passage in the Southern Ocean. *J. Hered.*, **99**:2 (2008), 137–48.

Iken, K., B. Bluhm, and K. Dunton, Benthic food-web structure serves as indicator of water mass properties in the southern Chukchi Sea. *Deep Sea Res. II*, **57** (2010), 71–85.

Iken, K., B. A. Bluhm, and R. Gradinger, Food web structure in the high Arctic Canada Basin: evidence from δC^{13} and δN^{15} analysis. *Polar Biol.*, **28**:3 (2005), 238–49.

IOPAN, 2010; Available from: www.iopan.gda.pl/biodaff/.

Jakobsson, M., A. Grantz, Y. Kristoffersen, and R. Macnab, Physiography and bathymetry of the Arctic Ocean. In R. Stein and R. MacDonald (eds.), *The Organic Carbon Cycle in the Arctic Ocean* (Berlin, Germany: Springer, 2004), pp. 1–6.

Janussen, D. and H. M. Reiswig, Hexactinellida (Porifera) from the ANDEEP III expedition to the Weddell Sea, Antarctica. *Zootaxa*, **2136** (2009), 1–20.

Junge, K., F. Imhoff, T. Staley, and J. W. Deming, Phylogenetic diversity of numerically important Arctic sea-ice bacteria cultured at subzero temperature. *Microb. Ecol.*, **43**:3 (2002), 315–28.

Kaiser, D., H. J. Griffiths, D. K. A. Barnes, S. N. Brandao, and A. Brandt, Is there a distinct continental slope fauna in the Antarctic? *Deep Sea Res. II* (in press).

Kaiser, S. and D. K. A. Barnes, Southern Ocean deep-sea biodiversity: sampling strategies and predicting responses to climate change. *Clim. Res.*, **37**:2–3 (2008), 165–79.

Kedra, M., M. Wlodarska-Kowalczuk, and J. M. Weslawski, Decadal change in macrobenthic soft-bottom community structure in a high Arctic fjord (Kongsfjorden, Svalbard). *Polar Biol.*, (2009), 1–11.

Kirchman, D. L., V. Hill, M. T. Cottrell, *et al.*, Standing stocks, production, and respiration of phytoplankton and heterotrophic bacteria in the western Arctic Ocean. *Deep Sea Res. II*, **56**:17 (2009), 1237–48.

Kock, K. -H., *Antarctic Fish and Fisheries* (Cambridge, UK: Cambridge University Press, 1992).

Korb, R. E., M. J. Whitehouse, and P. Ward, SeaWiFS in the Southern Ocean: spatial and temporal variability in phytoplankton biomass around South Georgia. *Deep Sea Res. II*, **51** (2004), 99–116.

Kosobokova, K., Seasonal variations in the vertical distribution and age composition of *Microcalanus pygmaeus*, *Oithona similis*, *Oncaea borealis* and *O. notopus* populations in the central Arctic basin. In *Biologiya Tsentral'nogo Arkicheskogo Basseyna* (Moscow: Nauka, 1980), pp. 167–82.

Kosobokova, K. and H. J. Hirche, Zooplankton distribution across the Lomonosov Ridge, Arctic Ocean: species inventory, biomass and vertical structure. *Deep Sea Res. I*, **47**:11 (2000), 2029–60.

Kotwicki, L. and F. Fiers, *Paracrenhydrosoma oceaniae*, new species, Harpacticoida, arctica, Fjord, Svalbard. *Annales Zoologici*, **55** (2005), 467–75.

Koubbi, P., M. Moteki, G. Duhamel, *et al.*, Ecological importance of micronektonic fish for the ecoregionalisation of the Indo-Pacific sector of the Southern Ocean: role of myctophids. *Deep Sea Res. II* (unpublished manuscript).

Kroncke, I., Macrofauna communities in the Amundsen Basin, at the Morris Jesup Rise and at the Yermak Plateau (Eurasian Arctic Ocean). *Polar Biol.*, **19**:6 (1998), 383–92.

Kuklinski, P. and P. D. Taylor, A new genus and some cryptic species of Arctic and boreal calloporid cheilostome bryozoans. *J. Mar. Biol. Assoc. UK*, **86**:5 (2006), 1035–46.

Kuklinski, P. and P. D. Taylor, Arctic species of the cheilostome bryozoan *Microporella*, with a redescription of the type species. *J. Nat. Hist.*, **42**:27–28 (2008), 1893–906.

Laakmann, S., M. Kochzius, and H. Auel, Ecological niches of Arctic deep-sea copepods: vertical partitioning, dietary preferences and different trophic levels minimize inter-specific competition. *Deep Sea Res. I*, **56**:5 (2009), 741–56.

Laws, R. M., The significance of vertebrates in the Antarctic marine ecosystem. In G. A. Llano (ed.), *Adaptations within Antarctic Ecosystem* (Houston: Gulf Publishing, 1977), pp. 411–38.

Le Romancer, M., M. Gaillard, C. Geslin, and D. Prieur, Viruses in extreme environments. *Rev. Environ. Sci. Biotechnol.*, **6**:1–3 (2007), 17–31.

Lecroq, B., A. J. Gooday, and J. Pawlowski, Global genetic homogeneity in the deep-sea foraminiferan *Epistominella exigua* (Rotaliida: Pseudoparrellidae). *Zootaxa*, **2096** (2009), 23–32.

Leese, F. and C. Held, Identification and characterization of microsatellites from the Antarctic isopod *Ceratoserolis trilobitoides*: nuclear evidence for cryptic species. *Conserv. Genet.*, **9**:5 (2008), 1369–72.

Loseto, L. L., G. A. Stern, D. Deibel, *et al.*, Linking mercury exposure to habitat and feeding behaviour in Beaufort Sea beluga whales. *J. Mar. Syst.*, **74**:3–4 (2008), 1012–24.

Lovejoy, C., R. Massana, and C. Pedros-Alio, Diversity and distribution of marine microbial eukaryotes in the Arctic Ocean and adjacent seas. *Appl. Environ. Microbiol.*, **72**:5 (2006), 3085–95.

MacDonald, I. R., B. A. Bluhm, K. Iken, S. Gagaev, and S. Strong, Benthic macrofauna and megafauna assemblages in the Arctic deep-sea Canada Basin. *Deep Sea Res. II*, **57** (2010), 136–52.

Makabe, R., H. Hattori, M. Sampei, *et al.*, Regional and seasonal variability of zooplankton collected using sediment traps in the southeastern Beaufort Sea, Canadian Arctic. *Polar Biol.*, (2009), 1–14.

Malyutina, M. and A. Brandt, Diversity and zoogeography of Antarctic deep-sea munnopsidae (Crustacea, Isopoda, Asellota). *Deep Sea Res. II*, **54** (2007), 1790–805.

Manes, S. S. and R. Gradinger, Small scale vertical gradients of Arctic ice algal photophysiological properties. *Photosynth. Res.*, **102**:1 (2009), 53–66.

Martin, J. H. and S. E. Fitzwater, Iron deficiency limits phytoplankton growth in the north-east Pacific subarctic. *Nature*, **331**:6154 (1988), 341–3.

Mecklenburg, C. W. and T. A. Mecklenburg, *Arctic Marine Fish Distribution and Taxonomy*. Minigrant report to Arctic Ocean Diversity Census of Marine Life Project, Auke Bay, Alaska (2008).

Mecklenburg, F., D. Stein, B. Sheiko, and N. Chernova, Russian-American long-term census of the Arctic: benthic fishes trawled in the Chukchi Sea and Bering Strait in 2004. *Northwest Nat.*, **83** (2007), 168–87.

Meredith, M. P. and J. C. King, Rapid climate change in the ocean west of the Antarctic Peninsula during the second half of the 20th century. *Geophys. Res. Lett.*, **32**:19 (2005).

Miller, D. G. and I. Hampton, Biology and ecology of the Antarctic krill (*Euphausia superba* Dana): a review. *BIOMASS Sci. Ser.*, **9** (1989), 1–66.

Moore, S. E., K. M. Stafford, and L. M. Munger, Acoustic and visual surveys for bowhead whales in the western Beaufort and far northeastern Chukchi seas. *Deep Sea Res. II*, **57** (2010), 153–7.

Mumm, N., On the summerly distribution of mesozooplankton in the Nansen Basin, Arctic Ocean. *Rep. Polar Res.*, **92** (1991), 1–173.

Naidu, A. S., Marine surficial sediments. Section 1.4 in *Bering, Chukchi and Beaufort Seas: Coastal and Ocean Zones Strategic Assessment Data Atlas* (Washington, DC: NOAA Strategic Assessment Branch, Ocean Assessment Division, 1988).

O'Driscoll, R. L., G. J. Macaulay, and S. Gauthier, Distribution, abundance and acoustic properties of Antarctic silverfish (*Pleuragramma antarcticum*) in the Ross Sea. *Deep Sea Res.* (in press).

O'Loughlin, P. M., G. Paulay, N. Davey, and F. Michonneau, Antarctic Region as a marine biodiversity hotspot for echinoderms: diversity and diversification of sea cucumbers. *Deep Sea Res. II* (in press).

Orr, J. C., V. J. Fabry, O. Aumont, *et al.*, Anthropogenic ocean acidification over the twenty-first century and its impact on calcifying organisms. *Nature*, **437**:7059 (2005), 681–6.

Page, T. J. and K. Linse, More evidence of speciation and dispersal across the Antarctic Polar Front through molecular systematics of Southern Ocean *Limatula* (Bivalvia: Limidae). *Polar Biol.*, **25**:11 (2002), 818–26.

Pasternak, A., E. Arashkevich, K. Tande, and T. Falkenhaug, Seasonal changes in feeding, gonad development and lipid stores in *Calanus finmarchicus* and *C. hyperboreus* from Malangen, northern Norway. *Mar. Biol.*, **138**:6 (2001), 1141–52.

Perovich, D. K., S. V. Nghiem, T. Markus, and A. Schweiger, Seasonal evolution and interannual variability of the local solar energy absorbed by the Arctic sea ice-ocean system. *J. Geophys. Res.*, **112**:C3 (2007).

Piepenburg, D., N. V. Chernova, C. F. Von Dorrien, *et al.*, Megabenthic communities in the waters around Svalbard. *Polar Biol.*, **16**:6 (1996), 431–46.

Pinchuk, A. I. and R. R. Hopcroft, Seasonal variations in the growth rates of euphausiids (*Thysanoessa inermis, T. spinifera*, and *Euphausia pacifica*) from the northern Gulf of Alaska. *Mar. Biol.*, **151**:1 (2007), 257–69.

Piraino, S., B. A. Bluhm, R. Gradinger, and F. Boero, *Sympagohydra tuuli* gen. nov and sp nov (Cnidaria: Hydrozoa) a cool hydroid from the Arctic sea ice. *J. Mar. Biol. Assoc. UK*, **88**:8 (2008), 1637–41.

Pomeroy, L. R., S. A. Macko, P. H. Ostrom, and J. Dunphy, The microbial food web in Arctic seawater: concentration of dissolved free amino acids and bacterial abundance and activity in the Arctic Ocean and in Resolute Passage. *Mar. Ecol. Prog. Ser.*, **61** (1990), 31–40.

Rapp, H. T., D. Janussen, and O. S. Tendal, Calcareous sponges from abyssal and bathyal depths in the Weddell Sea, Antarctica. *Deep Sea Res. II* (in press).

Raskoff, K. A., R. R. Hopcroft, K. N. Kosobokova, J. E. Purcell, and M. Youngbluth, Jellies under ice: ROV observations from the Arctic 2005 hidden ocean expedition. *Deep Sea Res. II*, **57** (2010), 111–26.

Raskoff, K. A., J. E. Purcell, and R. R. Hopcroft, Gelatinous zooplankton of the Arctic Ocean: in situ observations under the ice. *Polar Biol.*, **28**:3 (2005), 207–17.

Raupach, M. J., C. Held, and J. W. Wagele, Multiple colonization of the deep sea by the Asellota (Crustacea: Peracarida: Isopoda). *Deep Sea Res. II*, **51**:14–16 (2004), 1787–95.

Raupach, M. J., C. Mayer, M. Malyutina, and J. W. Wagele, Multiple origins of deep-sea Asellota (Crustacea: Isopoda) from shallow waters revealed by molecular data. *Proc. R. Soc. Lond. B Biol. Sci.*, **276**:1658 (2009), 799–808.

Raupach, M. J. and J. W. Wagele, Distinguishing cryptic species in Antarctic Asellota (Crustacea: Isopoda) – a preliminary study of mitochondrial DNA in *Acanthaspidia drygalskii. Antarct. Sci.*, **18**:2 (2006), 191–8.

Raymond, B. and G. Hosie, Network-based exploration and visualisation of ecological data. *Ecol. Model.*, **220**:5 (2009), 673–83.

Rodriguez, E., P. J. Lopez-Gonzalez, and M. Daly, New family of sea anemones (Actiniaria, Acontiaria) from deep polar seas. *Polar Biol.*, **32**:5 (2009), 703–17.

Rogacheva, A. V., Revision of the Arctic group of species of the family Elpidiidae (Elasipodida, Holothuroidea). *Mar. Biol. Res.*, **3**:6 (2007), 367–96.

Ronowicz, M., M. Wlodarska-Kowalczuk, and P. Kuklinski, Factors influencing hydroids (Cnidaria: Hydrozoa) biodiversity and distribution in Arctic kelp forest. *J. Mar. Biol. Assoc. UK*, **88**:8 (2008), 1567–75.

Rothrock, D. A., Y. Yu, and G. A. Maykut, Thinning of the Arctic sea-ice cover. *Geophys. Res. Lett.*, **26**:23 (1999), 3469–72.

Sakshaug, E., Primary and secondary production in the Arctic Seas. In R. Stein and R. MacDonald (eds.), *The Organic Carbon Cycle in the Arctic Ocean* (Berlin, Germany: Springer, 2004), pp. 57–82.

Scheidat, M., K.-H. Kock, A. Friedlaender, L. Lehnert, and R. Williams, *Preliminary results of aerial surveys around Elephant Island and the South Shetland Islands SC/59/IA21* (2007).

Scheidat, M., K.-H. Kock, A. Friedlaender, L. Lehnert, and R. Williams, *Using helicopters to survey Antarctic minke whale abundance in the ice SC/59/IA20* (2007).

Schrodl, M., J. M. Bohn, N. Brenke, E. Rolan, and E. Schwabe, Abundance, diversity and latitudinal gradients of south-eastern Atlantic and Antarctic abyssal gastropods. *Deep Sea Res. II* (in press).

Schwabe, E., J. M. Bohn, W. Engl, K. Linse, and M. Schrodl, Rich and rare – first insights into species diversity and abundance of Antarctic abyssal Gastropoda (Mollusca). *Deep Sea Res. II*, **54** (2007), 1831–47.

Serreze, M. C., M. M. Holland, and J. Stroeve, Perspectives on the Arctic's shrinking sea-ice cover. *Science*, **315**:5818 (2007), 1533–6.

Seuthe, L., G. Darnis, C. W. Riser, P. Wassmann, and L. Fortier, Winter-spring feeding and metabolism of Arctic copepods: insights from faecal pellet production and respiration measurements in the southeastern Beaufort Sea. *Polar Biol.*, **30**:4 (2007), 427–36.

Shaffer, S. A., Y. Tremblay, H. Weimerskirch, *et al.*, Migratory shearwaters integrate oceanic resources across the Pacific Ocean in an endless summer. *Proc. Natl. Acad. Sci. USA*, **103**:34 (2006), 12799–802.

Shaffer, S. A., H. Weimerskirch, D. Scott, *et al.*, Spatiotemporal habitat use by breeding sooty shearwaters *Puffinus griseus. Mar. Ecol. Prog. Ser.*, **391** (2009), 209–20.

Sirenko, B. I., *List of Species of Free-living Invertebrates of Eurasion Arctic Seas and Adjacent Deep Waters* (St. Petersburg: Russian Academy of Sciences, 2001).

Sirenko, B. I. and S. Y. Gagaev, Unusual abundance of macrobenthos and biological invasions in the Chukchi Sea. *Russ. J. Mar. Biol.*, **33**:6 (2007), 355–64.

Sirenko, B. I. and S. V. Vassilenko, Fauna and zoogeography of benthos of the Chukchi Sea. *Explor. Fauna Seas*, **61**:69 (2009), 1–230.

Smith, S. L. and S. Schnack-Schiel, Polar zooplankton. In W. O. J. Smith (ed.), *Polar Oceanography, Part B: Chemistry, Biology, and Geology* (San Diego: Academic Press, 1990), pp. 527–98.

Sogin, M. L., H. G. Morrison, J. A. Huber, *et al.*, Microbial diversity in the deep sea and the underexplored "rare biosphere." *Proc. Natl. Acad. Sci. USA*, **103** (2006), 12115–20.

Sohn, R. A., C. Willis, S. Humphris, *et al.*, Explosive volcanism on the ultraslow-spreading Gakkel Ridge, Arctic Ocean. *Nature*, **453**:7199 (2008), 1236–8.

Spalding, M. D., H. E. Fox, G. R. Allen, *et al.*, Marine ecoregions of the world: a bioregionalization of coastal and shelf areas. *Bioscience*, **57**:7 (2007), 573–83.

Stempniewicz, L., K. Blachowlak-Samolyk, and J. M. Weslawski, Impact of climate change on zooplankton communities, seabird populations and Arctic terrestrial ecosystem – a scenario. *Deep Sea Res. II*, **54**:23–26 (2007), 2934–45.

Stepanjants, S. D. and K. N. Kosobokova, Medusae of the genus *Rhabdoon* (Hydrozoa: Anthomedusae: Tubularioidea) in the Arctic Ocean. *Mar. Biol. Res.*, **2**:6 (2006), 388–97.

Strugnell, J. M., M. A. Collins, and A. L. Allcock, Molecular evolutionary relationships of the octopodid genus *Thaumeledone* (Cephalopoda: Octopodidae) from the Southern Ocean. *Antarct. Sci.*, **20**:3 (2008), 245–51.

Thomas, D. and G. Dieckmann (eds.), *Sea Ice: An Introduction to its Physics, Chemistry, Biology and Geology* (New York: Wiley-Blackwell, 2009).

d'Udekem d'Acoz, C. and W. Vader, On *Liljeborgia fissicornis* (M. Sars, 1858) and three related new species from Scandinavia, with a hypothesis on the origin of the group fissicornis. *J. Nat. Hist.*, **43**:33–34 (2009), 2087–139.

Vassilenko, S. V. and V. V. Petryashov (eds.), *Illustrated Keys to Free-living Invertebrates of Eurasian Arctic Seas and Adjacent Deep Waters, Vol. 1. Rotifera, Pycnogonida, Cirripedia, Leptostraca, Mysidacea, Hyperiidae, Caprellidea, Euphausiacea, Dendrobranchiata, Pleocyemata, Anomura, and Brachyura* (Alaska: University of Alaska Fairbanks, 2009).

Waller, C., D. K. A. Barnes, and P. Convey, Ecological contrasts across an Antarctic land-sea interface. *Austral. Ecol.*, **31** (2006), 656–66.

Walsh, J. E., Climate of the Arctic marine environment. *Ecol. Appl.*, **18**:Suppl.2 (2008).

Wassmann, P., M. Reigstad, T. Haug, *et al.*, Food webs and carbon flux in the Barents Sea. *Prog. Oceanogr.*, **71**:2–4 (2006), 232–87.

Whitehouse, M. J., M. P. Meredith, P. Rothery, *et al.*, Rapid warming of the ocean around South Georgia, Southern Ocean, during the 20th century: forcings, characteristics and implications for lower trophic levels. *Deep Sea Res. I*, **55**:10 (2008), 1218–28.

Wilson, N. G., M. Schrodl, and K. M. Halanych, Ocean barriers and glaciation: evidence for explosive radiation of mitochondrial lineages in the Antarctic sea slug *Doris kerguelenensis* (Mollusca, Nudibranchia). *Mol. Ecol.*, **18**:5 (2009), 965–84.

Wlodarska-Kowalczuk, M., Molluscs in Kongsfjorden (Spitsbergen, Svalbard): a species list and patterns of distribution and diversity. *Polar Res.*, **26**:1 (2007), 48–63.

Zuckerkandl, E. and L. B. Pauling, Molecular disease, evolution, and genetic heterogeneity. In M. Kasha and B. Pullman (eds.), *Horizons in Biochemistry* (New York: Academic Press, 1962), pp. 189–225.

Ocean life in motion

Many species of marine life – from microbes to whales – live their entire lives drifting or swimming, and may never see the seafloor or land. Sampling these nomads challenges the few scientists exploring the immensity of the ocean's more than 99% of the Earth's biosphere, and countless interconnected habitats. Much of the oceanic volume is far from shore and reaches depths without sunlight and with crushing pressure (Figure 7.1). Differences in temperature, salinity, pressure (depth), light, and oxygen and hence food determine where ocean life lives and why (Figure 7.2). Organisms move continuously in this fluid world – some drifting with the currents and others swimming in seasonal migrations over thousands of kilometers. Others migrate vertically more than a kilometer in a single day. For an organism only a few centimeters in length, this is not a commute, but a daily marathon.

Ocean life mostly travels to feed or reproduce. Their poetry of motion spans gentle pulsing of jellyfish or the powerful strokes of tuna and whales swimming across thousands of kilometers of blue ocean. But we don't know where they go, how they go, or why they go. Their movement complicates learning their diversity, distribution, and abundance. We will never sample the immense ocean completely. Nevertheless, new tools refined during the last decade helped us understand life in motion, demonstrating untold diversity of drifting microbes, blue highways in the ocean, oceanic globetrotters, and predator hot spots.

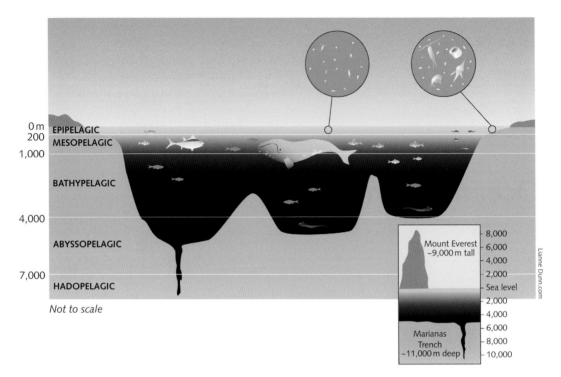

Figure 7.1 Ocean life in motion is defined by their waters.
This drawing shows pelagic ecosystems of Planet Ocean, where differences in light, temperature, salinity, and nutrients determine productivity. Productivity in turn limits who can live there.

Small is big

Their small size (Figure 7.3) belies the importance of marine microbes. They make half the food that fuels all life on Earth, and they control the global nitrogen, sulfur, iron, and manganese cycles. Without microbes, life on Earth would end. A single liter of seawater contains 100,000,000 to 1,000,000,000 individual microbes and without pyrosequencing a study is doing well to collect genetic sequences for more than 0.01%. One study estimated 20,000 microbial kinds (OTUs) in that liter of seawater. More recently, archaea alone were estimated at 3,000 OTUs per liter. So far molecular tools applied to protists project species richness comparable to the archaea. Microbes are the least known and physiologically most complex forms of life on Earth.

Mind-boggling microbial diversity makes it difficult to count. Counting a few dominant types in a sample is easy, but many rare forms complicate the job. At the rate of 20,000 kinds of microbes in one liter, the 1.35 trillion-billion liters in the ocean could hold unfathomable numbers. Many rare species were first found in hydrothermal vent fluid, but rare is now known to be common. In other words, rare microbes dominate microbial communities, and only a few species are abundant.

Beginning in 2004, the *International Census of Marine Microbes* (ICoMM) collected more than 1,200 samples spread among all major oceans, representing 583 bacterial, 120 archaeal, and 59 eukaryotic datasets and more than 25 million sequences. Importantly, because genetic tools are used to characterize microbial samples, the term "sequences" is the microbial vernacular for "count" and thus approximates the numbers of individuals sampled. Given the size and complexity of this dataset, ICoMM is only just beginning its analysis, but early findings are intriguing. Microbiologists estimate approximately 100,000,000,000,000,000,000,000,000,000 (or 10^{29}) total bacteria in the global ocean. Collectively microbes are up to 90% of total ocean biomass, and the remaining minority includes all the invertebrates, fish, and whales in the sea! There are differences among the bacteria, archaea, and protists that make up the microbial world; in some areas archaea are as abundant as bacteria, but only 10% as diverse. Protist diversity is similar to the archaea, but may be more diverse in some places.

How rare is rare? In other words, if a sample has many rare "species," does this mean that there are lots of rare forms that occur everywhere in low numbers ("everything is everywhere") or are rare forms just that? Different depths and seasons in the Arctic show similar biogeography of rare and common forms. It is possible sequencing picks up only the commonest of the rare or, more intriguingly, that rare forms live everywhere. Are rare species active or are they marginal, dormant forms that do very little? An analogy might be a stray tropical fish swept up to Nova Scotia in the Gulf Stream, but destined to die quickly. It is thus more a biogeographic footnote than ecologically important. Some rare species in the South Pacific are metabolically active in nutrient and other cycles, and rare species at hydrothermal vents bloom when environmental conditions are good.

As expected from past studies and in their global survey, ICoMM found the most abundant type of life on Earth, a group of Alphaproteobacteria, which cycle carbon, nitrogen, and sulfur, represent half of microbial abundance and 25% of microbial biomass. Alphaproteobacteria are everywhere.

The blue-green bacteria (Cyanobacteria) of the genera *Prochlorococcus* and *Synechococcus* live in sunlit waters at concentrations of up to 100,000

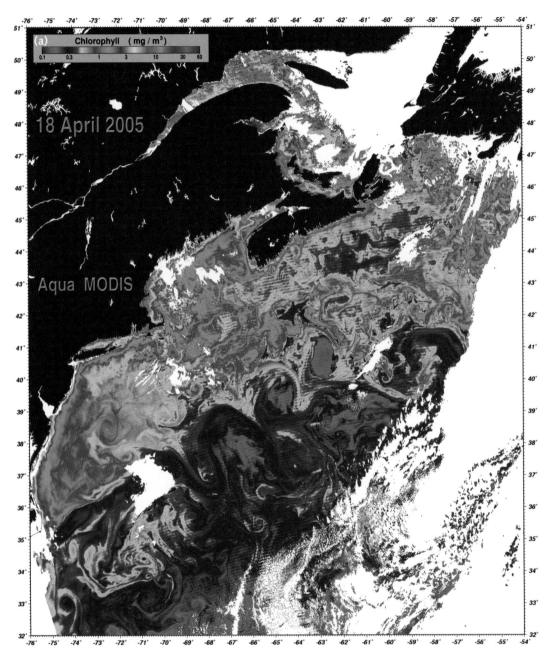

Figure 7.2 The complexity of ocean motion.
Satellite images of chlorophyll and temperature trace flowing surface waters off eastern North America. **(a)** This satellite false-color image of the Gulf Stream shows phytoplankton traveling curved highways and circular gyres in the water. The photosynthesis in phytoplankton feeds all marine life – except the peculiar communities near vents and seeps. In the photograph, the red color shows abundant phytoplankton in productive water and the blue and purple color shows the sparse phytoplankton in less productive water. **(b)** This second satellite image of the Gulf Stream shows a false-color image of surface temperatures. The water temperature indicated by colors in the image traces movements of Planet Ocean and visualizes the array of watery habitats.

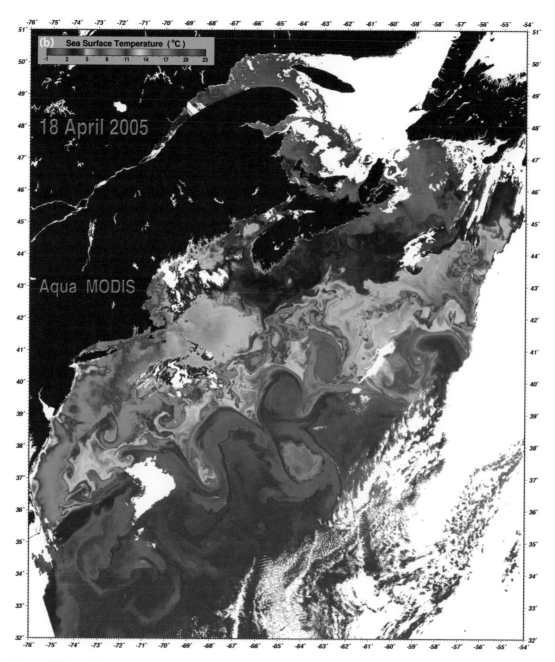

Figure 7.2 *(cont.)*

(a)

(b)

Figure 7.3 Ocean drifters.
Plankton shown here range from microbes measuring far less than the width of a human hair out to jellyfish almost as long as a baseball bat.
(a) *Protoperidinium pellucidum*, a colorized image of a protist dinoflagellate.
(b) Microbes from the Pacific Ocean off Southern California, stained to fluoresce. The numerous faint dots are viruses, and the brighter and less numerous medium dots are bacteria or archaea. The central object or dot is a photosynthetic protist with brightly stained nucleus.
(c) "Squidworm," an extraordinary worm found off Southeast Asia during a Census cruise by US and Filipino researchers and photographed by a *National Geographic* photographer.
(d) A bioluminescent jellyfish, *Aequorea macrodactyla*, photographed during a Census expedition to the Celebes Sea.
(e) *Neocalanus robustior*, an abundant Arctic copepod that eats phytoplankton, are key prey for fish.

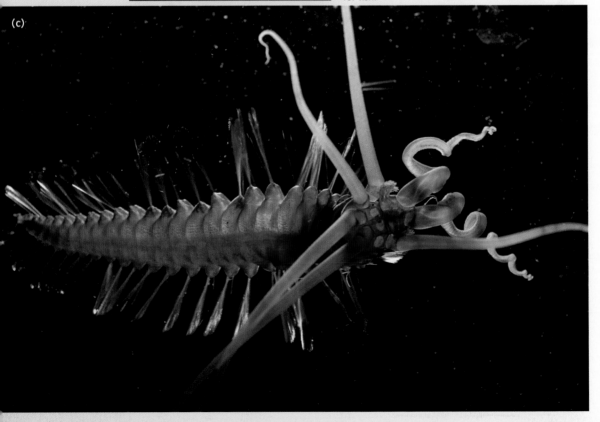

(c)

(d)

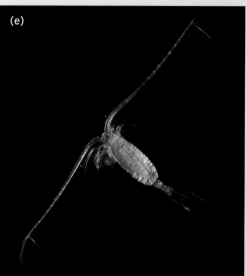

(e)

(f) The predatory *Rhizophysa* resembles a sprite dancing in the night sea. A diver captured it in the mid-Atlantic for DNA barcoding. This relative of the Man-o'-War shown in three views measures 10 centimeters long, but its stinging tentacles stretch 50 to 60 centimeters, when fishing for food in shallow water.

(f)

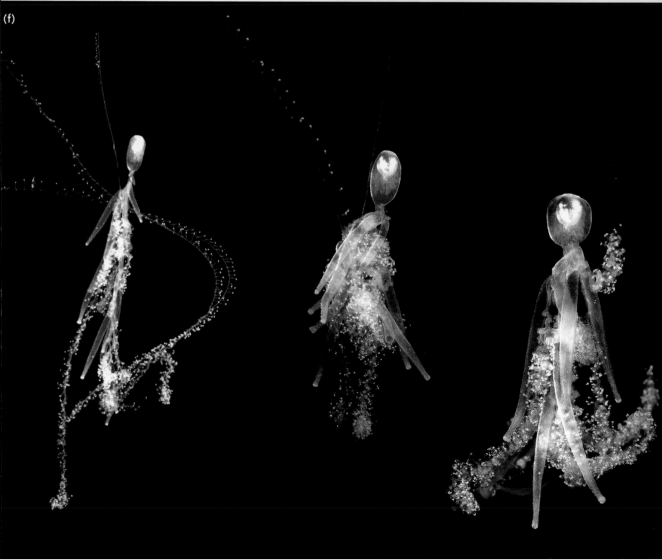

to 1,000,000 individuals per milliliter of seawater. Their photosynthesis produces perhaps half the primary production in the ocean. Cyanobacteria were the first organisms to produce oxygen as a byproduct of photosynthesis and oxygenate the Earth's atmosphere billions of years ago. We and most animals owe our existence to them. Among archaea, the phylum Crenarchaeota, once thought to live only in extreme environments, are now known to be widespread in the ocean and may outnumber bacteria below 100 meters, where they help cycle carbon. Among eukaryotes, abundant dinoflagellates include forms that cause harmful algal blooms. But dinoflagellates are notorious for having many copies of each gene, so are overcounted in studies that enumerate numbers of sequences. Tiny eukaryotes whose diversity has only been recognized within the last 10 years are most abundant. An unclassified dinoflagellate first reported from Antarctica can number 29,000 cells per liter. This microbe also occurs in the Arctic, Pacific, and Atlantic Oceans, Framvaren Fjord in Norway, and the Black Sea, but whether it is a single cosmopolitan species or multiple forms remains unknown.

Few microbes swim and it is tempting to assume they could disperse like the scent of holiday turkey through our homes and become present everywhere. But new work suggests microbial communities stay within discrete ocean volumes with distinct temperature and salinity called *water masses* (Figure 7.2). Many water masses span thousands of kilometers in distance and thousands of meters in depth. In the same water mass, microbes from an 8,000 kilometer stretch of deep water in the North Atlantic are similar over thousands of kilometers, but differ from microbes only hundreds of meters away in a separate water mass. This was true for common Proteobacteria and also for rare forms, suggesting that the rare ones did not stray in from elsewhere.

In stark contrast to this finding, in a sample of pyrosequencing analysis, the 20 most abundant kinds of Arctic and Antarctic bacteria were shared between poles. This pattern rests on just one genetic sequence, and we do not know yet how many of these occur at locations in between. Thus, true bipolar species may be as uncommon in microbes as in animals (see Chapter 6).

Although microbes play well-established roles in natural cycles, within the last 5 to 10 years researchers have found novel roles for them in cycling of carbon, nitrogen, sulfur, and iron. These new roles include how they use methane and ammonium in environments without oxygen, and the discovery of archaea that use energy in ammonia to build biological molecules.

These little powerhouses also play a key role in climate in that bacteria fix nitrogen gas from the atmosphere and ocean and convert it into a valuable nutrient for phytoplankton. Just a few kinds of bacteria fix nitrogen, and in laboratory experiments they react strongly to increased seawater acidity, a phenomenon already seen in the ocean. In similar experiments, seawater acidity changes metabolism and carbon uptake in *Prochlorococcus* and *Synechococcus*. If acidification continues, nitrogen and carbon cycles may change too, yet scientists scarcely know the roles and identities of key microbes.

The time traveler's microbe

Biologists would dearly love to ride H. G. Wells' *Time Machine* back in time to understand why the living world is what it is today. The extraction of DNA from preserved material to reproduce living dinosaurs in Stephen Spielberg's *Jurassic Park* is a stretch. Fossil DNA is rare and its interpretation controversial. But fat-related *lipids* are more robust than DNA and can persist in sediments for billions of years and help trace evolution. Though the puzzle evolves constantly and key pieces have disappeared over time, molecular analyses teach much about evolution. And much effort has gone into the lipid database. Like genetic databases, scientists around the world input data for future analyses. ICoMM began with an existing library, the Lipid Maps database (www.lipidmaps.org), which focused on biomedical applications, and broadened its scope to include marine lipids. With *MICROBIS*, the ICoMM microbial database, genetic and lipid data are placed in a precise geographic location that opens up a range of new analyses. For example, analysis of more than 100 types of *diatoms*, a group of phytoplankton, showed that rhizosolenoid diatoms evolved 91.5 million years ago, give or take 1.5 million years. Although not quite a time machine, this newly precise timing is novel, and it beats time travel and trying to elude the dinosaurs that ruled the Earth when rhizosolenoids evolved!

Lilliputian swimmers

Microbes are not the only drifters in the ocean. Small zooplankton swim and drift, although their small size and weak swimming make them planktonic (Figure 7.3). The *Census of Marine Zooplankton* (CMarZ) focused on zooplankton that live all of their lives in the water column, the *holoplankton*. Excluding the *meroplankton*, which are primarily larval stages of seafloor animals that spend a small portion of their life as drifting plankton, zooplankton encompass about 7,000 known species in 15 phyla. CMarZ targeted potential biodiversity hot spots throughout the world, including such remote regions as Southeast Asia, the polar oceans, and water below 5,000 meters. All these regions yielded new species to the Census, and all will yield more. CMarZ is the first global-scale synthesis of marine zooplankton to bring together all major zooplankton taxa. In every ocean basin during more than 90 CMarZ cruises since 2004, CMarZ filtered millions of cubic meters of water at more than 12,000 locations. An additional 6,500 archived samples supplemented these collections. From them, CMarZ mapped global distributions of protistan Foraminifera and Radiolaria, and wrote a new global monograph about Hydrozoa (jellyfish). They also examined basin-scale diversity of Atlantic Ocean ostracod crustaceans, and Indian Ocean chaetognath arrow worms, and will add others.

Why all of this effort? Besides the roles zooplankton play in food webs noted in Chapter 2, they also cycle carbon, nitrogen, and other ocean elements, though differently than microbes. In particular, they export food from surface waters down to layers beneath. Although the diversity of 7,000 species of zooplankton pales by more than a million next to global marine diversity (see Chapter 3), many zooplankton are abundant and widespread species.

While zooplankton are not diverse globally they can be diverse locally. A net collection from a single oceanic location could contain hundreds of copepod crustacean species, representing 10% of their global total. Imagine a sample containing this large a proportion of global biodiversity for any other group. Because zooplankton habitat is water itself, it is less complex than three-dimensional coral reefs or the microbial and chemical complexity of sediments. The bacterial world is defined by milliliter gradients in nutrients or oxygen, whereas swimming zooplankton experience a bigger world of thousands of cubic meters or less. For them, depth and neighboring organisms also define habitat. Consider the deep-sea jellyfish, *Pandea rubra*, that spends part of its life attached to a swimming mollusk.

Though many zooplankton are cosmopolitan or widely distributed, some look-alikes are genetically distinct species that require genetic barcoding to distinguish. As of late 2009, 2,000 of 7,000 described zooplankton species had already been barcoded, including protists, jellyfish, calanoid copepods, euphausiid krill, ostracods, pteropod mollusks, and chaetognath arrow worms. Barcoding revealed genetically distinct populations of copepods, krill, and arrow worms. CMarZ hopes to have most known zooplankton species barcoded by the completion of the first Census of Marine Life in October 2010.

These different strategies discovered 89 new species, of which 52 have already been formally described. These include new jellyfish from Norway, the Gulf of Maine, and the deep ocean, and at least 15 new species of ostracod crustaceans, new species of copepod, and shrimp-like mysid crustaceans. The Census added an additional 8% to the known copepods in Southeast Asian waters and 2% to the global total. In the open ocean CMarZ found a peak in species number at 750 meters and fewer species at higher latitudes. *Herbivores* that feed on phytoplankton dominate the upper layers and widely distributed carnivores and detritus feeders dominate at depth. Filtering large volumes of sparsely populated deep water with large, fine-meshed nets collected rare specimens from copepod crustaceans to fishes. The Census discovered a unique and diverse fauna under Arctic sea ice that earlier studies largely missed because they sampled in summer months after ice dynamics critical to Arctic and Antarctic ocean life had finished (see Chapter 6).

Endemism is rarer in deeper dwellers than shallower species. Frequent endemism in the Arctic and Antarctic contrasts with less diversity at these high latitudes. For example, only about 4% of global holozooplankton live in the Arctic, perhaps reflecting the difficulty of adapting to cold and ice.

The *Future of Marine Animal Populations* (FMAP) studied latitudinal patterns in drifting euphausiid crustaceans, foraminifera, different groups of fishes, and cephalopod mollusks. This work expands a previous FMAP study that showed a subtropical peak in diversity of billfish, tuna, and other predators that paralleled foraminiferan diversity and increased diversity in warmer water up to a point. Then it declined, perhaps because further warming reduced the solubility and thus availability of oxygen, another significant diversity predictor. Other metabolic issues, such as overheating in the warmest waters, may also come into play and explain the mid-latitude peak for oceanic animals that contrasted the equatorial peak for many coastal organisms (see Chapter 5).

Because zooplankton indicate environmental change, CMarZ has analyzed their patterns over decades of sampling. Since 1946, *Continuous*

Plankton Recorders on ships have surveyed the North Atlantic. By passively filtering out the organisms that pass across a revolving band of silk mesh as it is towed, this simple device has produced valuable biological datasets on the North Atlantic and elsewhere. They show that in the last few decades in the North Sea, warmer-water species have displaced cold-water copepods, a critical food resource for many larval fishes. These data have also demonstrated *"regime shifts,"* zooplankton changes related to shifts in water masses in the Northwest Atlantic, Northeast Pacific, and Northwest Pacific. Preserved samples in Japan showed decadal oscillations of copepod diversity and abundance over 40 years. Zooplankton abundances in South Africa's Benguela Current actually multiplied 100-fold over previous decades, possibly reflecting climate change. A rare 130-year plankton time series from the Black Sea showed increased comb jellies replaced copepods and reduced anchovy recruitment. Thus, knowledge of the ocean past has helped understand declining fisheries stocks and linked them to zooplankton. Researchers also anticipate Arctic warming will change the timing and magnitude of phytoplankton production and move more species through the Arctic.

From drifters to swimmers

Because biodiversity spans from genes to ecosystems, research can appropriately focus on single species rather than communities for some groups of organisms. With most species of large vertebrates, from fishes to whales, focusing on where they live and why is timely. On the one hand, sampling communities of microbes and zooplankton across the ocean at different times is necessary and workable. Sampling communities of whales, seabirds, and big fishes, on the other hand, is impractical and destructive, so researchers focus on observation of individuals.

Scientists have long marked organisms with tags and fluorescent particles, and photographed unique scars on whale tail flukes to learn where whales move. Knowing where animals move can define populations of commercial species and locate critical habitats. Although it may improve fishing efficiency by telling fishermen where the fish move, it may also show how populations are linked and could be protected or restored. It may eventually fill in black boxes on where species move.

A tale of two zip codes

In the Pacific Northwest of Canada and the United States, salmon is king. I am not referring to king (Chinook) salmon, but to the six species of salmon at the center of the history, culture, cuisine, economy, and politics of that part of the world. Artists admire salmon in Haida art and tourists buy smoked salmon as they leave Vancouver airport. Salmon research generates interest and controversy far beyond the biologists doing the research.

POST, the *Pacific Ocean Shelf Tracking Project*, inserted small radio tags into thousands of individuals spread over 16 species of fish and two species of squid, and then monitored their movement and survival with a "curtain" of listening receivers that detect them as they swim past (Figure 7.4). At first they tried to study the oceanic portion of the Pacific salmon life cycle. Do salmon travel specific areas of the ocean the way they follow specific rivers? Where and when do salmon die in the ocean? Pacific salmon species are *anadromous*, swimming into freshwater rivers and streams to spawn. They may stay there for two years before swimming back to the ocean. Salmon abundance and mortality in freshwater are well known, but in the larger and more open ocean, abundance and mortality were unknowable. Consider that a given *cohort*, or "generation," of salmon may move (and die) anywhere within a 16,000 kilometer loop as they move from Oregon to Alaska and on to Japan over a five-year stretch. *That* is a fisheries management nightmare!

The rapid expansion of human development in the North Pacific in the twentieth century increased agriculture, fishing, river damming, and logging, and climate is now changing. It is hardly surprising that some Pacific salmon populations dropped below 10% of their historical levels. These declines punctuate the need to know where individuals are lost from populations. An added complication is salmon aquaculture, which has created heated controversy about how hatchery-reared and, most especially, farmed salmon interact with natural salmon populations.

POST began with a "two zip code" theory that salmon live at one location in freshwater and another in the ocean, following earlier work that found salmon and their predators use different Pacific ocean areas seasonally. POST followed early successes with salmon by tagging squid and sturgeon.

When salmon swim past listening curtains and are detected (Figure 7.4), the information is clear, but when an individual is not detected it may be

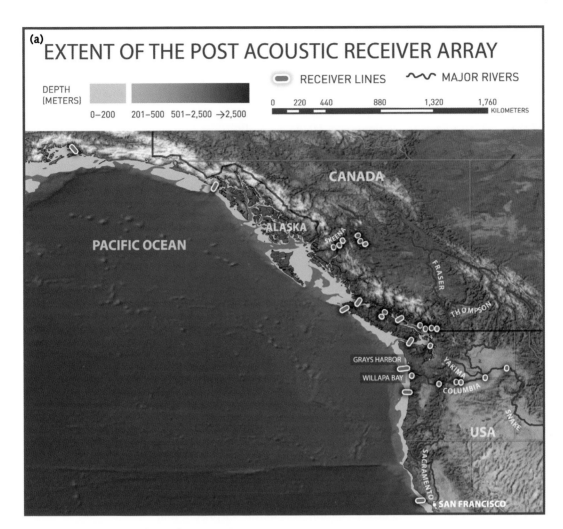

Figure 7.4 Checkpoints listen for migrants.
The Census arrayed lines of acoustic receivers on the bottom of rivers and along the Pacific continental shelf. As tagged fish with radio transmitters pass between receivers, they are identified and counted. **(a)** A map of Census receivers along the Pacific coast of North America. **(b)** A Census acoustic receiver displayed in Vancouver Aquarium's Strait of Georgia exhibit. Submerged receivers at "choke points" along the coast count and identify tagged fish as they pass by. **(c)** How an array of receivers "listens" for passing fish, whose tags emit radio waves (gray waves) that are detected by receivers (yellow) and relayed to a nearby ship.

(b)

(c)

Figure 7.4 (cont.)

because it died, swam past the line undetected, or its tag failed or was lost. Or perhaps the individual remained between listening curtains. POST curtains deployed at narrow points in migratory routes (river entrances, inland waterways, narrow straits) reduce these ambiguities, but do not encompass all habitats within the movement range.

Less than 5% of the *smolt*, or migrating freshwater juveniles, of anadromous rainbow trout (steelhead), *Oncorhynchus mykiss*, survive to adulthood, and this rate varies within the Pacific Northwest. This variability determines good and poor cohorts. POST found most individuals moved downstream daily less than 1 kilometer in smaller rivers but farther than 80 kilometers in larger rivers. Once in the ocean, populations migrated at similar speeds. Wild smolt survived migration better than hatchery-reared ones, which knew little of natural predators and suffered higher mortality soon after release.

During the 1990s, coho salmon, *Oncorhynchus kisutch*, almost disappeared from the Strait of Georgia off the coast of British Columbia and did not recover once the fishery closed. Warmer waters reduced southern populations and increased northern ones, but little was known about coho survival in the ocean. POST's acoustic tags showed inconsistent survival in the rivers in hatchery-reared salmon. In the ocean, mortality varied among populations as the migrating fish moved as far as 750 kilometers from their Oregon release. Survival was poor in summer, but better in September. As hatchery-released juveniles moved slowly downriver through the estuary, their physiology, behavior, and survival differed from wild populations. Coho survival was affected by season and by freshwater and marine habitats.

Dam building during the 1970s coincided with declines in chinook salmon, *Oncorhynchus tshawytscha*, in the Snake River in the northwestern United States. The Snake and Yakima Rivers flow into the Columbia River. Engineers have improved fish passages and spillways, and carried fish past impounded water. Despite high cost and improved smolt survival downriver, adult returns remained poor. Perhaps transport around the impound area reduced later survival in the coastal ocean. POST asked, what role do the dams play?

POST tagged 4,000 hatchery salmon and tracked them through rivers into the ocean, detecting five juveniles 2,500 kilometers away in their Alaskan listening array. They found that Yakima juveniles, which pass through four lower dams, were five times more likely to survive to adulthood than Snake River juveniles, which pass through eight dams. But survival in the lower river was similar for both populations, suggesting that survival was largely determined *after* they entered the ocean.

POST worked with an expensive program that transported juveniles around the Snake River dam system on barges. The fish survived during the barge trip itself, but barging did not improve survival from the last dam to the ocean. Also the barged juveniles did not return to the river at much greater rates than those that made their own way over the dams and down the river. Even though barged juveniles arrived at the ocean sooner, their mortality in the ocean was similar to those that made their own way in the river. Because barging did not improve survival, the dams may not be the culprits harming salmon. Some endangered Columbia and Snake River chinook and steelhead stocks downstream survive the dams in those rivers as well or better than the same species migrating from the dam-free Fraser River just to the north in British Columbia. Advocates of dam removal and related mitigations must scrutinize these experiments carefully for clues to improve survival.

White sturgeon, *Acipenser transmontanus*, as long as 6 meters and weighing more than 600 kilograms are the largest fish in freshwater in North America and in the world. Truly a freshwater Methuselah, it is a living fossil that has barely changed in over 65 million years and it can live longer than 150 years – except that most stocks in British Columbia are in trouble. Rapidly declining numbers of anadromous young that migrate to the ocean may cause the declines of 27% in Fraser River populations. To begin testing this hypothesis, POST helped tag 100 white sturgeon in the Fraser River in 2008, but results are still incomplete.

Farther south, green sturgeon, *Acipenser medirostris*, is in trouble too. Anadromous, like the related white sturgeon, they spawn only in three rivers and congregate in Oregon and Washington estuaries. Though caught along the coast from Baja California to the Bering Sea, their migration is unknown. Additional listening curtains from California to Washington augmented POST receivers, to match the sturgeon's range. Green sturgeon tagged north of Vancouver Island in winter migrated to rivers and estuaries in California, Oregon, and Washington in spring. Individual fish repeatedly used specific estuaries and rivers. This information can help recover this species.

Like many other sharks, predatory six-gill sharks, *Hexanchus griseus*, grow slowly and produce few offspring, which makes them vulnerable. POST tracked 59 young adults for as long as four years and found they moved no more than a few kilometers a day. During fall and winter their north to

south travel spanned only 8 kilometers, and expanded to only
120 kilometers northward in late spring and summer. Most of the
sharks remained in Puget Sound. In one year, 10% left the area, and as
some individuals matured, 46% strayed 200–350 kilometers to the north
and one strayed 1,400 kilometers south. The movements of mature six-gill
sharks remains unknown, but they have been caught kilometers from the
coast by longline fishermen. Six-gill sharks seem to lack the wanderlust
of most of the species tracked by Census tags.

Sockeye salmon, *Oncorhynchus nerka*, suffer great mortality. POST
singled this salmon species out as part of an interdisciplinary effort to
solve a complex puzzle. In the Fraser River alone, 150 distinct populations
move 100 to 1,200 kilometers upriver. Different populations move in pre-
dictable migrations in four clusters between June and October. In some
years, 50–95% of fall spawners, more than four million fish, perish during
the river migration, raising questions of sustainability.

Why would fish migrate in a season when migration is tantamount
to suicide? POST combined tracking with physiological measurements,
particularly concentrations of reproductive hormones and the degree to which
they were physiologically prepared to move into freshwater. For late-spawning
sockeye salmon, preparation for reproduction is the primary factor in driving
ocean migration. Elevated reproductive hormones in late spawners may
trigger early migrations before the primary spring/summer spawning
optimum.

The Census found that salmon that failed to reach the river to
spawn were more stressed, and less prepared for freshwater, than
those that succeeded. Necessary salts available in the ocean are scarce in
freshwater, hence the movement from salt to freshwater challenges migrants.
Indeed, this is why organisms such as echinoderms (sea stars, for example)
have been unable to invade freshwater over evolutionary time. They lack the
physiological capacity to cope with low salt concentrations. Fish that migrated
into freshwater before their gills were modified to deal with freshwater
frequently died, even when in good reproductive condition. Triggers in the
open ocean can hasten reproductive hormone development. Reproduction
drives late-season spawning sockeye migration. Elevated hormones in late-
season spawners may trigger migration before the spawning optimum.
If individuals become reproductively prepared too quickly, they may find
themselves trying to mate before they are ready, with dire consequences.
The analogy to the human species is almost irresistible!

Although some salmon die in rivers, multitudes die somewhere in the ocean. Larger acoustic arrays in deeper water and smaller tags may help to pinpoint graveyards in the ocean. Indeed, the *Ocean Tracking Network*, a new international research program led by Canada and spearheaded by Census scientist Ron O'Dor, will do just that.

Blue highways and truck stops

Tagging of Pacific Predators (TOPP) tagged and tracked large animals that swim far in the ocean and seabirds that fly above them. TOPP tagged mostly in the Pacific, where highly mobile species feed, reproduce, and play across the largest ocean on Earth. TOPP first assumed that several species would travel particular routes – ocean highways – to converge in ocean hot spots of abundance with favorable temperature and food. A group of about 90 scientists planned ambitious tracking of 23 predators at the top of the food web: whales and tunas, sharks and squid, seals and sea lions, turtles and seabirds. TOPP began tagging familiar species and added new ones as they improved technologies and methodologies. They developed a strategy of tracking *guilds*, species that live in a similar way, so they could share methods and compare animals with similar physiologies and habitat. Tagging 4,400 animals and 23 species, TOPP discovered truly globetrotting nomads like sooty shearwaters and stay-at-homes like female sea lions. White sharks migrated from coastal California to a meeting place the Census dubbed the "white shark café." Visiting the café off the coast of Hawaii and traveling back demonstrated strong and persistent homing behavior.

Some salmon sharks (*Lamna ditropis*) tagged in the Alaskan Arctic in the summer strayed as far south as the subtropical North Pacific in winter and spring. They spent the most time in water cooler than 10 °C above 50 meters, but some swam into both cold 2 °C water and warm 24 °C water. Physiological measurements showed cardiac and other adaptations to cold water where their fish prey abound.

Pacific bluefin tuna (*Thunnus orientalis*) and loggerhead turtles (*Caretta caretta*) mostly swim west to east, breeding in the western Pacific and then migrating as juveniles to the central California coast to feed. After spawning, tuna migrate into the South Pacific, encompassing the largest known home range among marine species. Leatherback turtles (*Dermochelys coriacea*), like loggerheads, breed on the beaches of

Figure 7.5 Seabirds enjoy endless summers.
The routes of sooty shearwaters from colonies in New Zealand extend from
the Southern Ocean to the Bering Sea as they migrate between hemispheres to enjoy
endless summers. The map traces 19 birds during breeding (light blue) and subsequent
migration. Yellow identifies the start of northward migration, and orange identifies
wintering grounds and southward migration.

Indonesia and migrate to California in late summer to feed on abundant
jellyfish. Traveling in the opposite direction, another population of leatherbacks
breeds on Costa Rican beaches in Central America and crosses the Pacific to
the food-poor subtropical Pacific. But the king of travel is the sooty shearwater
(*Puffinus griseus*), which travels the Pacific from the Antarctic to Alaska's
Bering Sea in an "endless summer" with abundant prey to eat and
favorable habitats for breeding (Figure 7.5).

As predicted, tagged juvenile Pacific bluefin tuna swam seasonally from
the western to the eastern Pacific region and then from Baja, California to
Oregon and congregated in a hot spot in the California Current.

Animals that TOPP equipped with electronic tags revealed their hot spots in the North Pacific. Predators of many species, from leatherback and loggerhead turtles to sooty shearwaters, to whales, tunas, seals, and sharks, all congregate at several locations in the California Current. There, seasonal upwelling of nutrients creates a rich supply of food that nourishes abundant prey for many top predators.

If the California Current is a cluster of hot spots, what are the "blue highways" of nomadic migrations? The North Pacific Transition Zone is one major highway and a hot spot too. Along this highway east to west across the Pacific, mixing subarctic and subtropical water creates a patchy, "trail of crumbs" across the ocean that is followed by tunas, seals and sea turtles, albatrosses and shearwaters. Some travel the highway, whereas for others it is a destination. Year after year, the females of two elephant seal populations that breed along a 1,200 kilometer span of the Mexican and California coastlines feed along the highway. Although the southern population must travel far, they travel it anyway. Individual white sharks repeatedly follow distinct highways to use their coastal hot spots again and again.

TOPP found out how animals divide up these highways and hot spots. Because the hot spots lie along highways we may think of them as "truck stops." Though many share the same routes and truck stops, they often use them differently so they don't compete. Pacific bluefin, yellowfin, and albacore tuna are related, but the *endothermic* bluefin tuna maintain their body temperature and can tolerate cold. Because yellowfin and albacore tuna, and mako and blue sharks lack this ability, their body temperatures drop in cold water, which confines them to tropical to mid-latitude waters. Blue whales travel 100–200 kilometers off the coast of California to feed on euphausiid krill, whereas humpbacks chase schools of small fish inshore. Black-footed albatrosses feed in warm productive waters, while the cooler and patchier waters to the north favor Laysan albatrosses. Animals also partition habitat vertically. Thus, bluefin feed on abundant sardine and anchovy near the surface layer, whereas albacore feed in deeper water.

Animals also avoid competing in hot spots by eating different diets. White sharks dine on marine mammals that congregate in coastal colonies, while mako sharks hunt fish on the continental shelf. Even sharks are not mindless eating machines. Salmon sharks feed on pollock and herring, but mako and blue sharks feed on squid and sardine.

Three hot spots within the California Current add seasonal separation. Tunas, blue, mako, thresher, and juvenile white sharks feed and grow in the Southern California Bight, and then tunas visit the California Marine Sanctuaries and Baja California seasonally. In short, species once thought to wander indiscriminately have well-defined "kitchens," "bedrooms," "hallways," and "nurseries."

TOPP discovered that unusual ocean conditions change the behavior in some species. In winter, sea lions usually stay close to shore, taking excursions for less than a day. When seasonal upwelling of nutrient-rich waters was delayed, however, they sought food 300–500 kilometers offshore and changed their diet from squid and anchovy to sardine and rockfish.

Managers of endangered species, protected areas, and fisheries can use knowledge of hot spots of diversity and activity in pelagic waters of the open sea. If we know where animals are likely to be, we can plan accordingly. Others want to know what lives in the deep blue sea.

BIBLIOGRAPHY

Adams, P. B., C. Grimes, J. E. Hightower, *et al.*, Population status of North American green sturgeon, *Acipenser medirostris. Environ Biol. Fishes*, **79**:3–4 (2007), 339–56.

Amaral-Zettler, L., L. F. Artigas, J. Baross, *et al.*, A global census of marine microbes. In A. D. McIntyre (ed.), *Life in the World's Oceans: Diversity, Distribution, and Abundance* (Oxford: Blackwell Publishing Ltd., 2010), pp. 223–45.

Amaral-Zettler, L. A., E. A. McCliment, H. W. Ducklow, and S. M. Huse, A method for studying protistan diversity using massively parallel sequencing of V9 hypervariable regions of small-subunit ribosomal RNA genes. *PLoS ONE*, **4**:7 (2009).

Anderson, S. D., B. H. Becker, and S. G. Allen, Observations and prey of white sharks, *Carcharodon carcharias*, at Point Reyes National Seashore: 1982–2004. *Calif. Fish. Game.*, **94**:1 (2008), 33–43.

Andrews, K. S., P. S. Levin, S. L. Katz, *et al.*, Acoustic monitoring of sixgill shark movements in Puget Sound: evidence for localized movement. *Can. J. Zool.*, **85**:11 (2007), 1136–42.

Angel, M. V., What is the deep sea? In D. J. Randall and A. P. Farrell (eds.), *Deep-sea Fishes* (San Diego: Academic Press, 1997), pp. 1–41.

Angel, M. V., The pelagic environment of the open ocean. In P. Tyler (ed.), *Ecosystems of the Deep Ocean* (Amsterdam: Elsevier, 2003), pp. 39–79.

Angel, M. V., *Atlas of Atlantic Planktonic Ostracods* (London: Natural History Museum, 2008).

Angel, M. V. and K. Blachowiak-Samolyk, Ostracods. *Rep. Polar Mar. Res.*, **592** (2009), 29–31.

Angel, M. V., K. Blachowiak-Samolyk, I. Drapun, and R. Castillo, Changes in the composition of planktonic ostracod populations across a range of latitudes in the North-east Atlantic. *Prog. Oceanogr.*, **73**:1 (2007), 60–78.

Angel, M. V., L. Nigro, and A. Bucklin, DNA barcoding of oceanic planktonic ostracoda: species recognition and discovery (World Conference on Marine Biodiversity, Valencia, Spain, 2008).

Arndt, C. E. and K. M. Swadling, Crustacea in Arctic and Antarctic sea ice: distribution, diet and life history strategies. *Adv. Mar. Biol.*, **51** (2006), 197–315.

Baum, J. K., R. A. Myers, D. G. Kehler, *et al.*, Collapse and conservation of shark populations in the Northwest Atlantic. *Science*, **299**:5605 (2003), 389–92.

Beamish, R. J., D. J. Noakes, G. A. McFarlane, *et al.*, The regime concept and natural trends in the production of Pacific salmon. *Can. J. Fish. Aquat. Sci.*, **56**:3 (1999), 516–26.

Beamish, R. J., R. M. Sweeting, K. L. Lange, and C. M. Neville, Changes in the population ecology of hatchery and wild coho salmon in the Strait of Georgia. *Trans. Am. Fish. Soc.*, **137**:2 (2008), 503–20.

Beaugrand, G., P. C. Reid, F. Ibañez, J. A. Lindley, and M. Edwards, Reorganization of North Atlantic marine copepod biodiversity and climate. *Science*, **296**:5573 (2002), 1692–4.

Blanco-Bercial, L., F. Alvarez-Marques, and A. Bucklin, Global phylogeographies of the planktonic copepod Clausocalanus based on DNA barcodes (Third International Conference for the Barcode of Life, Mexico City, Mexico, 2009).

Block, B. A., Physiological ecology in the 21st century: advancements in biologging science. *Integr. Comp. Biol.*, **45**:2 (2005), 305–20.

Block, B. A., D. P. Costa, G. W. Boehlert, and R. E. Kochevar, Revealing pelagic habitat use: the tagging of Pacific pelagics program. *Oceanol. Acta*, **25**:5 (2002), 255–66.

Block, B. A., D. P. Costa, and S. J. Bograd, A view of the ocean from Pacific predators. In A. D. McIntyre (ed.), *Life in the World's Oceans: Diversity, Distribution, and Abundance* (Oxford: Blackwell Publishing Ltd., 2010), pp. 291–311.

Bluhm, B. A. and R. Gradinger, Regional variability in food availability for Arctic marine mammals. *Ecol. Appl.*, **18**:2 (2008), S77–S96.

Bluhm, B. A., R. Gradinger, and S. Schnack-Schiel, Sea ice meio- and macrofauna. In D. Thomas and G. Dieckmann (eds.), *Sea Ice: An Introduction to its Physics, Chemistry, Biology and Geology* (Hoboken, NJ: Wiley-Blackwell, 2010), pp. 357–94.

Boltovskoy, D., N. Correa, and A. Boltovskoy, Marine zooplanktonic diversity: a view from the South Atlantic. *Oceanol. Acta*, **25**:5 (2002), 271–8.

Boltovskoy, D., N. Correa, and A. Boltovskoy, Diversity and endemism in cold waters of the South Atlantic: contrasting patterns in the plankton and the benthos. *Sci. Mar.*, **69** (2005), 17–26.

Bouillion, J., C. Gravili, F. Pages, J. M. Gili, and F. Boero (eds.), *An Introduction to Hydrozoa* (Paris: Memoires du Museum d'Histoire Naturelle, 2006).

Boustany, A. M., S. F. Davis, P. Pyle, *et al.*, Satellite tagging – expanded niche for white sharks. *Nature*, **415**:6867 (2002), 35–6.

Boustany, A. M., R. Matteson, M. R. Castleton, C. J. Farwell, and B. A. Block, Movements of Pacific bluefin tuna (*Thunnus orientalis*) in the Eastern North Pacific revealed with archival tags. *Prog. Oceanogr*, **86** (2010), 94–104.

Bradford, M. J. and J. R. Irvine, Land use, fishing, climate change, and the decline of Thompson River, British Columbia, coho salmon. *Can. J. Fish. Aquat. Sci.*, **57**:1 (2000), 13–6.

Brazelton, W. J., M. L. Sogin, and J. A. Baross, Multiple scales of diversification with natural populations of Archaea in hydrothermal chimney biofilms. *Environ. Microbiol. Rep.*, **2** (2009), 236–42.

Brinton, E., M. D. Ohman, A. W. Townsend, M. D. Knight, and A. L. Bridgeman, *Euphausiids of the World Ocean* (Berlin: Springer Verlag, 2000).

Bucklin, A., B. W. Frost, J. Bradford-Grieve, L. D. Allen, and N. J. Copley, Molecular systematic and phylogenetic assessment of 34 calanoid copepod species of the Calanidae and Clausocalanidae. *Mar. Biol.*, **142**:2 (2003), 333–43.

Bucklin, A., T. C. LaJeunesse, E.Curry, J. Wallinga, and K. Garrison, Molecular diversity of the copepod, *Nannocalanus minor*: genetic evidence of species and population structure in the North Atlantic Ocean. *J. Mar. Res.*, **54**:2 (1996), 285–310.

Bucklin, A. C., S. Nishida, S. Schnack-Schiel, *et al.*, A census of zooplankton of the global ocean. In A. D. McIntyre (ed.), *Life in the World's Oceans: Diversity, Distribution, and Abundance* (Oxford: Blackwell Publishing Ltd., 2010), pp. 247–65.

Bucklin, A., P. H. Wiebe, S. B. Smolenack, *et al.*, DNA barcodes for species identification of euphausiids (Euphausiacea, Crustacea). *J. Plankton Res.*, **29**:6 (2007), 483–93.

Buitenhuis, E., C. Le Quere, O. Aumont, *et al.*, Biogeochemical fluxes through mesozooplankton. *Global Biogeochem. Cycles*, **20**:2 (2006).

Chittenden, C., K. Butterworth, K. Cubitt, *et al.*, Maximum tag to body size ratios for an endangered coho salmon (*O. kisutch*) stock based on physiology and performance. *Environ. Biol. Fishes*, **84**:1 (2009), 129–40.

Chittenden, C. M., S. Sura, K. G. Butterworth, *et al.*, Riverine, estuarine and marine migratory behaviour and physiology of wild and hatchery-reared coho salmon *Oncorhynchus kisutch* (Walbaum) smolts descending the Campbell River, BC, Canada. *J. Fish Biol.*, **72**:3 (2008), 614–28.

Clapham, P. J., Humpback whale, *Megaptera novaeangliae*. In W. F. Perrin, B. Wursig, and J. G. M. Thewissen (eds.), *Encyclopedia of Marine Mammals* (New York: Elsevier-Academic Press, 2009), pp. 582–5.

Conway, D. V. P., R. G. White, J. Hugues-Dit-Ciles, C. P. Gallienne, and D. B. Robins, *Guide to the Coastal and Surface Zooplankton of the South-western Indian Ocean* (Plymouth: Marine Biological Association of the United Kingdom, 2003).

Cooke, S. J., S. G. Hinch, G. T. Crossin, *et al.*, Physiology of individual late-run Fraser River sockeye salmon (*Oncorhynchus nerka*) sampled in the ocean correlates with fate during spawning migration. *Can. J. Fish. Aquat. Sci.*, **63**:7 (2006), 1469–80.

Cooke, S. J., S. G. Hinch, A. P. Farrell, *et al.*, Abnormal migration timing and high en route mortality of sockeye salmon in the Fraser River, British Columbia. *Fisheries*, **29**:2 (2004), 22–33.

Coolen, M. J. L., G. Muyzer, W. I. C. Rijpstra, *et al.*, Combined DNA and lipid analyses of sediments reveal changes in Holocene haptophyte and diatom populations in an Antarctic lake. *Earth Planet Sci. Lett.*, **223**:1–2 (2004), 225–39.

Coronado, C. and R. Hilborn, Spatial and temporal factors affecting survival in coho salmon (*Oncorhynchus kisutch*) in the Pacific Northwest. *Can. J. Fish. Aquat. Sci.*, **55**:9 (1998), 2067–77.

Costa, D. P. and B. Sinervo, Field physiology: physiological insights from animals in nature. *Annu. Rev. Physiol.*, **66** (2004), 209–38.

Crossin, G. T., S. G. Hinch, D. W. Welch, *et al.*, Physiological profiles of sockeye salmon in the Northeastern Pacific Ocean and the effects of exogenous GnRH and testosterone on rates of homeward migration. *Mar. Freshw. Behav. Physiol.*, **42**:2 (2009), 89–108.

Damste, J. S. S., G. Muyzer, B. Abbas, *et al.*, The rise of the rhizosolenid diatoms. *Science*, **304**:5670 (2004), 584–7.

Damste, J. S. S., M. Strous, W. I. C. Rijpstra, *et al.*, Linearly concatenated cyclobutane lipids form a dense bacterial membrane. *Nature*, **419**:6908 (2002), 708–12.

De Vargas, C., M. Bonzon, N. W. Rees, J. Pawlowski, and L. Zaninetti, A molecular approach to biodiversity and biogeography in the planktonic foraminifer *Globigerinella siphonifera* (d'Orbigny). *Mar. Micropaleontol.*, **45**:2 (2002), 101–16.

Deibel, D. and K. L. Daly, Zooplankton processes in Arctic and Antarctic polynyas. In W. O. J. Smith and D. G. Barber (eds.), *Arctic and Antarctic Polynyas* (Elsevier, 2007), pp. 271–322.

Delong, E. F., Archaea in coastal marine environments. *Proc. Natl. Acad. Sci. USA*, **89**:12 (1992), 5685–9.

Diez, B., C. Pedros-Alio, and R. Massana, Study of genetic diversity of eukaryotic picoplankton in different oceanic regions by small-subunit rRNA gene cloning and sequencing. *Appl. Environ. Microbiol.*, **67**:7 (2001), 2932–41.

Edwards, M., D. G. Johns, G. Beaugrand, *et al.*, *Ecological Status Report: Results from the CPR Survey 2006/2007* (Plymouth: Sir Alister Hardy Foundation for Ocean Science Report, 2008).

Edwards, M., D. G. Johns, P. Licandro, A. W. G. John, and D. P. Stevens, *Ecological Status Report: Results from the CPR Survey 2005/2006* (Plymouth, UK, 2007).

Erickson, D. L. and M. A. H. Webb, Spawning periodicity, spawning migration, and size at maturity of green sturgeon, *Acipenser medirostris*, in the Rogue River, Oregon. *Environ. Biol. Fishes*, **79**:3–4 (2007), 255–68.

Fahy, E., S. Subramaniam, H. A. Brown, *et al.*, A comprehensive classification system for lipids. *Eur. J. Lipid Sci. Technol.*, **107**:5 (2005), 337–64.

Fahy, E., S. Subramaniam, R. C. Murphy, *et al.*, Update of the LIPID MAPS comprehensive classification system for lipids. *J. Lipid Res.*, **50** (2009), S9–S14.

Fiedler, P. C., S. B. Reilly, R. P. Hewitt, *et al.*, Blue whale habitat and prey in the California Channel Islands. *Deep Sea Res. II*, **45**:8–9 (1998), 1781–801.

Field, C. B., M. J. Behrenfeld, J. T. Randerson, and P. Falkowski, Primary production of the biosphere: integrating terrestrial and oceanic components. *Science*, **281**:5374 (1998), 237–40.

Fish, S. A., T. J. Shepherd, T. J. McGenity, and W. D. Grant, Recovery of 16S ribosomal RNA gene fragments from ancient halite. *Nature*, **417**:6887 (2002), 432–6.

Fu, F. X., M. E. Warner, Y. H. Zhang, Y. Y. Feng, and D. A. Hutchins, Effects of increased temperature and CO_2 on photosynthesis, growth, and elemental ratios in marine *Synechococcus* and *Prochlorococcus* (Cyanobacteria). *J. Phycol.*, **43**:3 (2007), 485–96.

Fuhrman, J. A., T. D. Sleeter, C. A. Carlson, and L. M. Proctor, Dominance of bacterial biomass in the Sargasso Sea and its ecological implications. *Mar. Ecol. Prog. Ser.*, **57**:3 (1989), 207–17.

Galand, P. E., E. O. Casamayor, D. L. Kirchman, and C. Lovejoy, Ecology of the rare microbial biosphere of the Arctic Ocean. *Proc. Natl. Acad. Sci. USA*, **106** (2009), 22427–32.

Gast, R. J., D. M. Moran, D. J. Beaudoin, *et al.*, Abundance of a novel dinoflagellate phylotype in the Ross Sea, Antarctica. *J. Phycol.*, **42**:1 (2006), 233–42.

Gavrilova, N. and J. R. Dolan, A note on species lists and ecosystem shifts: Black Sea tintinnids, ciliates of the microzooplankton. *Acta Protozool.*, **46**:4 (2007), 279–88.

Giovannoni, S. J., T. B. Britschgi, C. L. Moyer, and K. G. Field, Genetic diversity in Sargasso Sea bacterioplankton. *Nature*, **345**:6270 (1990), 60–3.

Goetze, E., Cryptic speciation on the high seas: global phylogenetics of the copepod family Eucalanidae. *Proc. R. Soc. Lond. B Biol. Sci.*, **270**:1531 (2003), 2321–31.

Goldman, K. J. and J. A. Musick, The biology and ecology of the salmon shark, *Lamna ditropis*. *Fish. Aquat. Resour. Ser.*, **13** (2008), 95–104.

Gorsky, G. and R. Fenaux, The role of Appendicularia in marine food chains. In Q. Bone (ed.), *The Biology of Pelagic Tunicates* (New York: Oxford University Press, 1998), pp. 161–9.

Grebmeier, J. M., L. W. Cooper, H. M. Feder, and B. I. Sirenko, Ecosystem dynamics of the Pacific-influenced Northern Bering and Chukchi Seas in the Amerasian Arctic. *Prog. Oceanogr.*, **71**:2–4 (2006), 331–61.

Gresh, T., J. Lichatowich, and P. Schoonmaker, An estimation of historic and current levels of salmon production in the Northeast Pacific ecosystem: evidence of a nutrient deficit in the freshwater systems of the Pacific Northwest. *Fisheries*, **25**:1 (2000), 15–21.

Hamasaki, K., A. Taniguchi, Y. Tada, R. A. Long, and F. Azam, Actively growing bacteria in the Inland Sea of Japan, identified by combined bromodeoxyuridine immunocapture and denaturing gradient gel electrophoresis. *Appl. Environ. Microbiol.*, **73**:9 (2007), 2787–98.

Hare, S. R., N. J. Mantua, and R. C. Francis, Inverse production regimes: Alaska and West Coast Pacific salmon. *Fisheries*, **24**:1 (1999), 6–14.

Herndl, G. J., T. Reinthaler, E. Teira, *et al.*, Contribution of Archaea to total prokaryotic production in the deep Atlantic Ocean. *Appl. Environ. Microbiol.*, **71**:5 (2005), 2303–9.

Heywood, J. L., M. V. Zubkov, G. A. Tarran, B. M. Fuchs, and P. M. Holligan, Prokaryoplankton standing stocks in oligotrophic gyre and equatorial provinces of the Atlantic Ocean: evaluation of inter-annual variability. *Deep Sea Res. II*, **53**:14–16 (2006), 1530–47.

Hosia, A. and F. Pages, Unexpected new species of deep-water Hydroidomedusae from Korsfjorden, Norway. *Mar. Biol.*, **151**:1 (2007), 177–84.

Huber, J. A., D. B. Mark Welch, H. G. Morrison, *et al.*, Microbial population structures in the deep marine biosphere. *Science*, **318**:5847 (2007), 97–100.

Hulbert, L. B., A. M. Aires-Da-Silva, V. F. Gallucci, and J. S. Rice, Seasonal foraging movements and migratory patterns of female *Lamna ditropis* tagged in Prince William Sound, Alaska. *J. Fish Biol.*, **67**:2 (2005), 490–509.

Hutchins, D. A., P. N. Sedwick, G. R. DiTullio, *et al.*, Control of phytoplankton growth by iron and silicic acid availability in the subantarctic Southern Ocean: experimental results from the SAZ Project. *J. Geophys. Res.*, **106**:C12 (2001), 31559–72.

Hyrenbach, K. D., P. Fernandez, and D. J. Anderson, Oceanographic habitats of two sympatric North Pacific albatrosses during the breeding season. *Mar. Ecol. Prog. Ser.*, **233** (2002), 283–301.

IPCC, *Climate Change 2007: Synthesis Report Contribution of Working Groups I, II and III to the Fourth Assessment Report of the Intergovernmental Panel on Climate Change* (Geneva, Switzerland: IPCC. 2007).

Jennings, R. M., A. Bucklin, and H. Ossenbrügger, Analysis of genetic diversity of planktonic gastropods from several ocean regions using DNA barcodes. *Deep Sea Res. II*, (in press).

Jennings, R. M., A. Bucklin, and A. Pierrot-Bults, Barcoding of arrow worms (Phylum Chaetognatha) from three oceans: genetic diversity and evolution within an enigmatic phylum. *PLoS ONE*, **5**:4 (2010), e9949, doi: 10.1371/journal.pone.0009949.

Jereb, P. and C. F. E. Roper, *Cephalopods of the World. An Annotated and Illustrated Catalogue of Species Known to Date* (Rome: FAO, 2005).

Jorgensen, S. J., J. Reeb, T. Chapple, *et al.*, Philopatry and migration of Pacific white sharks. *Proc. R. Soc. Lond. B Biol. Sci. B*, **277** (2010), 679–88.

Kane, J., Zooplankton abundance trends on Georges Bank, 1977–2004. *ICES J. Mar. Sci.*, **64**:5 (2007), 909–19.

Kappes, M. A., S. A. Shaffer, Y. Tremblay, *et al.*, Hawaiian albatrosses track interannual variability of marine habitats in the North Pacific. *Prog. Oceanogr.*, **86** (2010), 246–60.

Keiper, C. A., D. G. Ainley, S. G. Allen, and J. T. Harvey, Marine mammal occurrence and ocean climate off central California, 1986 to 1994 and 1997 to 1999. *Mar. Ecol. Prog. Ser.*, **289** (2005), 285–306.

Kiko, R., J. Michels, E. Mizdalski, S. B. Schnack-Schiel, and I. Werner, Living conditions, abundance and composition of the metazoan fauna in surface and sub-ice layers in pack ice of the western Weddell Sea during late spring. *Deep Sea Res. II*, **55**:8–9 (2008), 1000–14.

Kirchman, D. L. (ed.), *Microbial Ecology of the Oceans* (Hoboken: John Wiley & Sons, Inc., 2008).

Kitamura, M., D. J. Lindsay, and H. Miyake, Description of a new midwater medusa, *Tiaropsidium shinkaii* n. sp. (Leptomedusae, Tiaropsidae). *Plankton Biol. Ecol.*, **52**:2 (2005), 100–6.

Knittel, K. and A. Boetius, Anaerobic oxidation of methane: progress with an unknown process. *Annu. Rev. Microbiol.*, **63** (2009), 311–34.

Knowlton, N., R. E. Brainard, R. Fisher, *et al.*, Coral reef biodiversity. In A. D. McIntyre (ed.), *Life in the World's Oceans: Diversity, Distribution, and Abundance* (Oxford: Blackwell Publishing Ltd., 2010), pp. 65–77.

Konneke, M., A. E. Bernhard, J. R. de la Torre, *et al.*, Isolation of an autotrophic ammoniaoxidizing marine archaeon. *Nature*, **437**:7058 (2005), 543–6.

Kosobokova, K. and H. J. Hirche, Zooplankton distribution across the Lomonosov Ridge, Arctic Ocean: species inventory, biomass and vertical structure. *Deep Sea Res. I*, **47**:11 (2000), 2029–60.

Kuypers, M. M. M., A. O.Sliekers, G. Lavik, *et al.*, Anaerobic ammonium oxidation by anammox bacteria in the Black Sea. *Nature*, **422**:6932 (2003), 608–11.

Lavaniegos, B. E. and M. D. Ohman, Long-term changes in pelagic tunicates of the California Current. *Deep Sea Res. II*, **50**:14–16 (2003), 2473–98.

Li, W. K. W., Primary production of prochlorophytes, cyanobacteria, and eukaryotic ultraphytoplankton – measurements from flow cytometric sorting. *Limnol. Oceanogr.*, **39**:1 (1994), 169–75.

Lindley, S. T., M. L. Moser, D. L. Erickson, *et al.*, Marine migration of North American green sturgeon. *Trans. Am. Fish Soc.*, **137**:1 (2008), 182–94.

Lindsay, D. J. and J. C. Hunt, Biodiversity in midwater cnidarians and ctenophores: submersible-based results from deep-water bays in the Japan Sea and North-western Pacific. *J. Mar. Biol. Assoc. UK*, **85**:3 (2005), 503–17.

Lindsay, D., F. Pages, J. Corbera, *et al.*, The anthomedusan fauna of the Japan Trench: preliminary results from in situ surveys with manned and unmanned vehicles. *J. Mar. Biol. Assoc. UK*, **88**:8 (2008), 1519–39.

Lindsay, D. J. and I. Takeuchi, Associations in the deep-sea benthopelagic zone: the amphipod crustacean *Caprella subtilis* (Amphipoda: Caprellidae) and the holothurian *Ellipinion kumai* (Elasipodida: Elpidiidae). *Sci. Mar.*, **72**:3 (2008), 519–26.

Link, J. S., J. K. T. Brodziak, S. F. Edwards, *et al.*, Marine ecosystem assessment in a fisheries management context. *Can. J. Fish. Aquat. Sci.*, **59**:9 (2002), 1429–40.

Lopez, S., R. Melendez, and P. Barria, Feeding of the shortfin mako shark *Isurus oxyrinchus* Rafinesque, 1810 (Lamniformes: Lamnidae) in the Southeastern Pacific. *Rev. Biol. Mar. Oceanogr.*, **44**:2 (2009), 439–51.

Lopez-Garcia, P., F. Rodriguez-Valera, C. Pedros-Alio, and D. Moreira, Unexpected diversity of small eukaryotes in deep-sea Antarctic plankton. *Nature*, **409**:6820 (2001), 603–7.

Machida, R. J., M. U. Miya, M. Nishida, and S. Nishida, Molecular phylogeny and evolution of the pelagic copepod genus *Neocalanus* (Crustacea: Copepoda). *Mar. Biol.*, **148**:5 (2006), 1071–9.

Machida, R. J. and S. Nishida, Amplified fragment length polymorphism analysis of the mesopelagic copepods *Disseta palumbii* in the equatorial western Pacific and adjacent waters: role of marginal seas for genetic isolation of mesopelagic animals. *Deep Sea Res. II* (in press).

Mackas, D. L., S. Batten, and M. Trudel, Effects on zooplankton of a warmer ocean: recent evidence from the Northeast Pacific. *Prog. Oceanogr.*, **75**:2 (2007), 223–52.

McKinnell, S., J. J. Pella, and M. L. Dahlberg, Population-specific aggregations of steelhead trout (*Oncorhynchus mykiss*) in the North Pacific Ocean. *Can. J. Fish. Aquat. Sci.*, **54**:10 (1997), 2368–76.

Miller, K., S. Li, A. Schulze, *et al.*, *Late-run Gene Array Research* (Vancouver, BC: Pacific Salmon Commission Southern Boundary Restoration and Enhancement Fund, 2007).

Moon-van der Staay, S. Y., R. De Wachter, and D. Vaulot, Oceanic 18S rDNA sequences from picoplankton reveal unsuspected eukaryotic diversity. *Nature*, **409**:6820 (2001), 607–10.

Morard, R., F. Quillevere, G. Escarguel, *et al.*, Morphological recognition of cryptic species in the planktonic foraminifer *Orbulina universa*. *Mar. Micropaleontol.*, **71**:3–4 (2009), 148–65.

Moyle, P. B., *Inland Fishes of California* (Berkeley, CA: University of California Press, 2002).

Muir, W. D., S. G. Smith, J. G. Williams, E. E. Hockersmith, and J. R. Skalski, Survival estimates for migrant yearling chinook salmon and steelhead tagged with passive integrated transponders in the Lower Snake and Lower Columbia Rivers, 1993–1998. *N. Am. J. Fish. Manage.*, **21**:2 (2001), 269–82.

Murano, M. and K. Fukuoka, *A Systematic Study of the Genus Siriella (Crustacea: Mysida) from the Pacific and Indian Oceans, with Description of Fifteen New Species. Natural Museum of Nature and Science Monographs 36* (2008), 173 pp.

Myers, R. A., S. A. Levin, R. Lande, *et al.*, Hatcheries and endangered salmon. *Science*, **303**:5666 (2004), 1980.

Nair, V. R., S. U. Panampunnayil, H. U. K. Pillai, and R. Gireesh, Two new species of chaetognatha from the Andaman Sea, Indian Ocean. *Mar. Biol. Res.*, **4**:3 (2008), 208–14.

Nelson, T., W. J. Gazey, and K. K. English, *Status of White Sturgeon in the Lower Fraser River: Report on the Findings of the Lower Fraser River White Sturgeon Monitoring and Assessment Program 2007* (Vancouver, BC: Fraser River Sturgeon Conservation Society. 2008).

Nishida, S. and N. Cho, A new species of *Tortanus* (Atortus) (Copepoda, Calanoida, Tortanidae) from the coastal waters of Nha Trang, Vietnam. *Crustaceana*, **78** (2005), 223–35.

Ohtsuka, S., S. Nishida, and R. J. Machida, Systematics and zoogeography of the deep-sea hyperbenthic family Arietellidae (Copepoda: Calanoida) collected from the Sulu Sea. *J. Nat. Hist.*, **39**:27 (2005), 2483–514.

Ohtsuka, S., A. Tanimura, R. J. Machida, and S. Nishida, Bipolar and antitropical distributions of planktonic copepods. *Fossils*, **85** (2009), 6–13.

O'Malley, M. A., "Everything is everywhere: but the environment selects": ubiquitous distribution and ecological determinism in microbial biogeography. *Stud. Hist. Philos. Sci. C*, **39**:3 (2008), 314–25.

Ortman, B. D., DNA barcoding the medusozoa and ctenophora. PhD thesis, University of Connecticut (2008).

Pace, N. R., A molecular view of microbial diversity and the biosphere. *Science*, **276**:5313 (1997), 734–40.

Pages, F., J. Corbera, and D. Lindsay, Piggybacking pycnogonids and parasitic narcomedusae on *Pandea rubra* (Anthomedusae, Pandeidae). *Plankton Benthos Res.*, **2**:2 (2007), 83–90.

Pages, F., P. Flood, and M. Youngbluth, Gelatinous zooplankton net-collected in the Gulf of Maine and adjacent submarine canyons: new species, new family (Jeanbouilloniidae), taxonomic remarks and some parasites. *Sci. Mar.*, **70**:3 (2006), 363–79.

Park, E. T. and F. D. Ferrari, Species diversity and distributions of pelagic calanoid copepods from the Southern Ocean. In I. Krupnik, M. A. Lang, and S. E. Miller (eds.), *Smithsonian at the Poles: Contributions to the International Polar Year Science* (Washington, DC: Smithsonian Institution Scholarly Press, 2008), pp. 143–80.

Payne, J., K. Andrews, C. Chittenden, *et al.*, Tracking fish movements and survival on the Northeast Pacific Shelf. In A. D. McIntyre (ed.), *Life in the World's Oceans: Diversity, Distribution, and Abundance* (Oxford: Blackwell Publishing Ltd., 2010), pp. 269–90.

Pearcy, W. G., *Ocean Ecology of North Pacific Salmonids* (Seattle, WA: University of Washington Press, 1992).

Peckham, S. H., D. M. Diaz, A. Walli, *et al.*, Small-scale fisheries bycatch jeopardizes endangered Pacific loggerhead turtles. *PLoS ONE*, **2**:10 (2007), e1041.

Pedros-Alio, C., Marine microbial diversity: can it be determined? *Trends Microbiol.*, **14**:6 (2006), 257–63.

Peijnenburg, K. T. C. A., J. A. J. Breeuwer, A. C. Pierrot-Bults, and S. B. J. Menken, Phylogeography of the planktonic chaetognath *Sagitta setosa* reveals isolation in European seas. *Evolution*, **58**:7 (2004), 1472–87.

Pershing, A. J., C. H. Greene, J. W. Jossi, *et al.*, Interdecadal variability in the Gulf of Maine zooplankton community, with potential impacts on fish recruitment. *ICES J. Mar. Sci.*, **62**:7 (2005), 1511–23.

Porter, A. D., D. Welch, E. Rechisky, *et al.*, *Pacific Ocean Shelf Tracking Project (POST): Results from the Acoustic Tracking Study on Survival of Columbia River Salmon, 2008* (2009). Report to the Bonneville Power Administration by Kitama Research Corporation, Contract No. 2003-11-00, Grant No. 00021107.

Raghoebarsing, A. A., A. Pol, K. T. van de Pas-Schoonen, *et al.*, A microbial consortium couples anaerobic methane oxidation to denitrification. *Nature*, **440**:7086 (2006), 918–21.

Rechisky, E. and D. Welch, Surgical implantation of acoustic tags: influence of tag loss and tag-induced mortality on free-ranging and hatchery-held spring Chinook (*O. tschawytscha*) smolts. In K. S. Wolf and J. S. O'Neal (eds.), *Tagging, Telemetry and Marking Measures for Monitoring Fish Populations: A Compendium of New and Recent Science for Use in Informing Technique and Decision Modalities* (Duvall, WA: The Pacific Northwest Aquatic Monitoring Partnership and Kwa Ecological Sciences, Inc., 2009).

Reid, P. C., J. M. Colebrook, J. B. L. Matthews, and J. Aiken, The Continuous Plankton Recorder: concepts and history, from plankton indicator to undulating recorders. *Prog. Oceanogr.*, **58**:2–4 (2003), 117–73.

Roden, G. I., Subarctic-subtropical transition zone of the North Pacific large-scale aspects and mesoscale structure. In J. A. Wetherall (ed.), *Biology, Oceanography, and Fisheries of the North Pacific Transition Zone and Subarctic Frontal Zone* (Washington, DC: Department of Commerce, 1991).

Roemmich, D. and J. McGowan, Climatic warming and the decline of zooplankton in the California Current. *Science*, **267**:5202 (1995), 1324–6.

Rutherford, S., S. D'Hondt, and W. Prell, Environmental controls on the geographic distribution of zooplankton diversity. *Nature*, **400**:6746 (1999), 749–53.

Scanlan, D. J., M. Ostrowski, S. Mazard, *et al.*, Ecological genomics of marine picocyanobacteria. *Microbiol. Mol. Biol. Rev.*, **73**:2 (2009), 249–99.

Schaller, H. A., P. Wilson, S. Haeseker, *et al.*, *Comparative Survival Study (CSS) of pit-Tagged Spring/Summer Chinook and Steelhead in the Columbia River Basin: Ten-year Retrospective Analyses Report.* Comparative Survival Study Oversight Committee and Fish Passage Center. 2007.

Schnack-Schiel, S. B., Aspects of the study of the life cycles of Antarctic copepods. *Hydrobiologia*, **453–454** (2001), 9–24.

Schnack-Schiel, S. B., J. Michels, E. Mizdalski, M. P. Schodlok, and M. Schroder, Composition and community structure of zooplankton in the sea ice-covered western Weddell Sea in spring 2004 – with emphasis on calanoid copepods. *Deep Sea Res. II*, **55**:8–9 (2008), 1040–55.

Scott, W. B. and E. J. Crossman, Freshwater fishes of Canada. *Bull. Fish. Res. Board Can.*, **184** (1973).

Sears, R. and W. F. Perrin, Blue whale, *Balaenoptera musculus*. In W. F. Perrin, B. Wursig, and J. G. M. Thewissen (eds.), *Encyclopedia of Marine Mammals* (New York: Elsevier-Academic Press, 2009), pp. 120–4.

Shaffer, S. A., Y. Tremblay, H. Weimerskirch, *et al.*, Migratory shearwaters integrate oceanic resources across the Pacific Ocean in an endless summer. *Proc. Natl. Acad. Sci. USA*, **103**:34 (2006), 12799–802.

Shaffer, S. A., H. Weimerskirch, D. Scott, *et al.*, Spatiotemporal habitat use by breeding sooty shearwaters *Puffinus griseus*. *Mar. Ecol. Prog. Ser.*, **391** (2009), 209–20.

Shillinger, G. L., D. M. Palacios, H. Bailey, *et al.*, Persistent leatherback turtle migrations present opportunities for conservation. *PLoS Biol.*, **6**:7 (2008), 1408–16.

Sirenko, B. I., *List of Species of Free-living Invertebrates of Eurasian Arctic Seas and Adjacent Deep Waters* (St. Petersburg: Russian Academy of Sciences, 2001).

Smith, S. L. and S. Schnack-Schiel, Polar zooplankton. In W. O. J. Smith (ed.), *Polar Oceanography, Part B: Chemistry, Biology, and Geology* (San Diego: Academic Press, 1990), pp. 527–98.

Sogin, M. L., H. G. Morrison, J. A. Huber, *et al.*, Microbial diversity in the deep sea and the underexplored "rare biosphere." *Proc. Natl. Acad. Sci. USA*, **103** (2006), 12115–20.

Strous, M., J. A. Fuerst, E. H. M. Kramer, *et al.*, Missing lithotroph identified as new planctomycete. *Nature*, **400**:6743 (1999), 446–9.

Sugisaki, H., Studies of long-term variation of ocean ecosystem/climate interactions based on the Odate collection: outline of the Odate Project. *PICES Press*, **14** (2006), 12–5.

Sydeman, W. J., R. D. Brodeur, C. B. Grimes, A. S. Bychkov, and S. McKinnell, Marine habitat "hotspots" and their use by migratory species and top predators in the North Pacific Ocean: introduction. *Deep Sea Res. II*, **53**:3–4 (2006), 247–9.

Tian, Y. J., H. Kidokoro, T. Watanabe, and N. Iguichi, The late 1980s regime shift in the ecosystem of Tsushima warm current in the Japan/East Sea: evidence from historical data and possible mechanisms. *Prog. Oceanogr.*, **77**:2–3 (2008), 127–45.

Tittensor, D. P., C. Mora, W. Jetz, *et al.*, Global patterns and predictors of marine biodiversity across taxa. *Nature* (2010) doi: 10.1038/nature09329.

Venter, J. C., K. Remington, J. F. Heidelberg, *et al.*, Environmental genome shotgun sequencing of the Sargasso Sea. *Science*, **304**:5667 (2004), 66–74.

Verheye, H. M., Decadal-scale trends across several marine trophic levels in the southern Benguela upwelling system off South Africa. *Ambio*, **29**:1 (2000), 30–4.

Verheye, H. M. and A. J. Richardson, Long-term increase in crustacean zooplankton abundance in the southern Benguela upwelling region (1951–1996): bottom-up or top-down control? *ICES J. Mar. Sci.*, **55**:4 (1998), 803–7.

Welch, D. W., E. L. Rechisky, M. C. Melnychuk, *et al.*, Survival of migrating salmon smolts in large rivers with and without dams. *PLoS Biol.*, **6**:12 (2008), 2940.

Weise, M. J., D. P. Costa, and R. M. Kudela, Movement and diving behavior of male California sea lion (*Zalophus californianus*) during anomalous oceanographic conditions of 2005 compared to those of 2004. *Geophys. Res. Lett.*, **33**:22 (2006).

Wells, H. G., *The Time Machine* (UK: William Heinemann, 1895).

Weng, K. C., A. M. Boustany, P. Pyle, *et al.*, Migration and habitat of white sharks (*Carcharodon carcharias*) in the eastern Pacific Ocean. *Mar. Biol.*, **152**:4 (2007), 877–94.

Weng, K. C., P. C. Castilho, J. M. Morrissette, *et al.*, Satellite tagging and cardiac physiology reveal niche expansion in salmon sharks. *Science*, **310**:5745 (2005), 104–6.

Whitman, W. B., D. C. Coleman, and W. J. Wiebe, Prokaryotes: the unseen majority. *Proc. Natl. Acad. Sci. USA*, **95**:12 (1998), 6578–83.

Worm, B., H. K. Lotze, and R. A. Myers, Predator diversity hotspots in the blue ocean. *Proc. Natl. Acad. Sci. USA*, **100**:17 (2003), 9884–8.

Wuchter, C., S. Schouten, H. T. S. Boschker, and J. S. S. Damste, Bicarbonate uptake by marine Crenarchaeota. *FEMS Microbiol. Lett.*, **219**:2 (2003), 203–7.

Zhu, F., R. Massana, F. Not, D. Marie, and D. Vaulot, Mapping of picoeucaryotes in marine ecosystems with quantitative PCR of the 18S rRNA gene. *FEMS Microbiol. Ecol.*, **52**:1 (2005), 79–92.

8

Into the deep

"*Provehito in Altum*," the motto of my undergraduate university and current employer, translates to "Launch Forth into the Deep." Glimpses of the dark solitude of the deep ocean create a lasting memory. Early morning in January, 1988, the research ship *Atlantis II* held position off Mexico. Another scientist and I, along with the pilot, climbed into *Alvin*, the world's best-known submersible. Like stuffing three people into the trunk of a Volvo, the other scientist and I shoehorned our modest frames into *Alvin*'s sphere much more easily than the 6'5" pilot. We all avoided tea or coffee because tiny *Alvin*'s toilet was a plastic bottle. None minded these discomforts because adventure and discovery lay ahead. As we sank to the seafloor through two thousand meters and stared out through the plate-sized portholes, light faded to total darkness interrupted only by flashes of bioluminescence from mid-water organisms. Inside, *Alvin*'s titanium hull protected us as dimly lit safety dials showed increasing depth and crushing pressure.

At the bottom of the sea

After sinking steadily through blackness for an hour, the outside lights showed an eerie landscape of rolling hills covered by mud. During his historic 1934 bathysphere dive to about 1,000 meters off the coast of Bermuda, explorer William Beebe described a similar lifeless plain, unaware the sediment hid a riot of species of Lilliputian life crawling around in it (Figure 8.1).

(a)

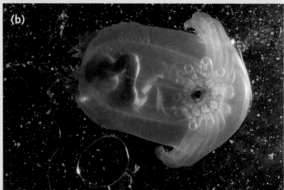

(b)

Figure 8.1 Deeper than light.
A diversity of life, from tiny crustaceans to rare fish, lives in the dark deep sea.
(a) Abyssochrysidae are a family of rare tropical snails that live in deep water. Like most deep-sea species, the Abyssochrysidae are smaller than their relatives that live in shallow water.
(b) The swimming sea cucumber, *Enypniastes*, differs from its sluggish relatives that spend their lives crawling along the seafloor.
(c) A new species of squat lobster photographed on the seafloor near Southwest Australia. The Census produced a new catalog of squat lobsters around the world.

(c)

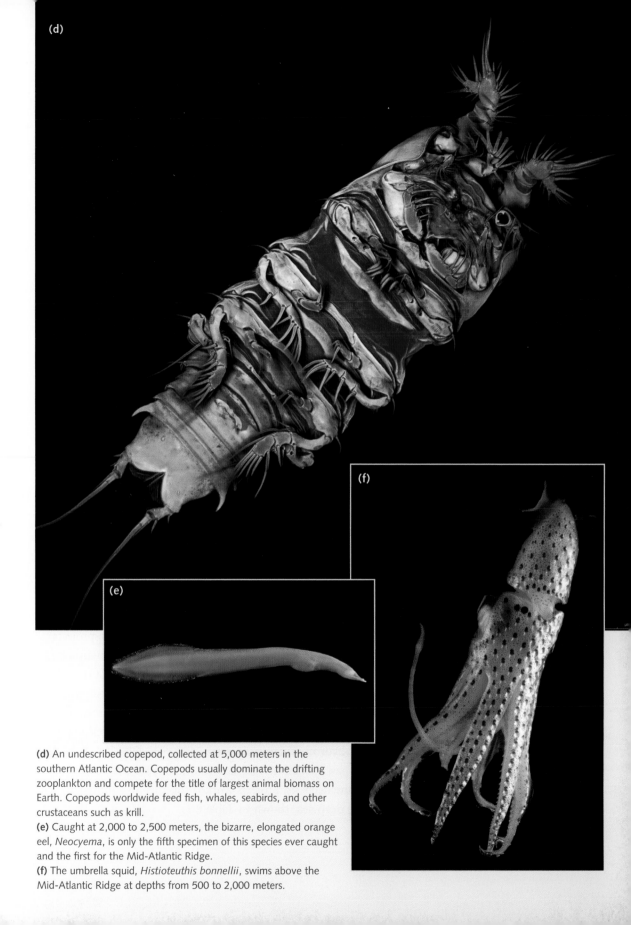

(d) An undescribed copepod, collected at 5,000 meters in the southern Atlantic Ocean. Copepods usually dominate the drifting zooplankton and compete for the title of largest animal biomass on Earth. Copepods worldwide feed fish, whales, seabirds, and other crustaceans such as krill.

(e) Caught at 2,000 to 2,500 meters, the bizarre, elongated orange eel, *Neocyema*, is only the fifth specimen of this species ever caught and the first for the Mid-Atlantic Ridge.

(f) The umbrella squid, *Histioteuthis bonnellii*, swims above the Mid-Atlantic Ridge at depths from 500 to 2,000 meters.

Table 8.1 The top ten reasons why the deep sea is special

10. We cannot extrapolate knowledge from other environments to the deep sea because of its special characteristics and dominant organisms.

9. The interconnected deep ocean and its large size make taxonomic coordination difficult, but vital to understand broad-scale patterns.

8. As the most pristine marine environment left on Earth, it is one of the few places where we can study natural processes where human impact is minimal.

7. Because they are so distant and often in international waters, the management and conservation of the deep sea is particularly complex and undeveloped.

6. The human footprint is creeping deeper into the ocean with unknown consequences because of many rare species, special habitats, and slow-growing species that produce few offspring.

5. Deep-sea habitats include some of the most unique and spectacular habitats on Earth, including some that changed our views of how ecosystems work.

4. The rate of discovery of abyssal nematode worms and abyssal plains species shows no sign of leveling off, illustrating the magnitude of the unknown.

3. The deep sea covers more of Earth than all other habitats combined, yet total sampling covers only a few football fields in area. Surely we must do better!

2. Just about every deep-sea sample has new species in it, and in some areas unknown species in a sample can greatly outnumber the known.

1. Entirely new types of deep-sea habitats have been found steadily over the last 100 years and new discoveries don't seem to be slowing down.

Now in 1988, *Alvin's* pilot radioed up to the surface, announcing we were on the bottom, and requested direction and distance to find the Guaymas hydrothermal vents. As the vents appeared through the blackness I thought that the spectacular photographs and video of Guaymas and other vents can scarcely capture the eye-popping smoking chimneys and beautiful bouquets of life clustered around them.

More than six decades after Beebe's first look into the deep sea, Census scientists dedicated five of their 14 field projects to its study. These projects focused on the continental slopes (COMARGE), chemosynthetic environments (ChEss), seamounts (CenSeam), the Mid-Atlantic Ridge (MAR-ECO), and the abyssal plains (CeDAMar). In crushing pressure and in darkness without photosynthesis, the deep sea is the largest frontier on Earth today. From the vast plains of the abyss with their sparse fauna, to the diverse continental margins with their spectacular hydrothermal vents, seeps, and rich diversity of species (Figure 8.1), the deep sea yielded myriad new knowledge for the Census. Census leaders from the deep-sea projects worked together to summarize the importance of the deep sea, shown in Table 8.1.

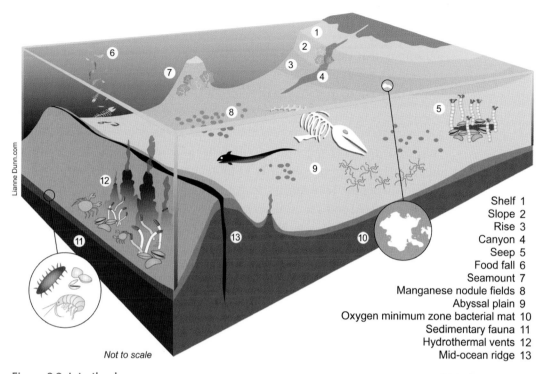

Lianne Dunn.com

Not to scale

Shelf 1
Slope 2
Rise 3
Canyon 4
Seep 5
Food fall 6
Seamount 7
Manganese nodule fields 8
Abyssal plain 9
Oxygen minimum zone bacterial mat 10
Sedimentary fauna 11
Hydrothermal vents 12
Mid-ocean ridge 13

Figure 8.2 Into the deep.
The drawing shows the array of specialized habitats on the seafloor, thousands of meters below the surface. The Census standardized datasets to enable comparisons among these specialized habitats.

The *continental margins* extend from the shoreline to the abyssal plains, but the continental slopes from the shelf edge to the abyssal plain form the landward portion of the deep sea (Figure 8.2). On the slopes, most food sinks from surface waters or arrives horizontally from adjacent shelves. Scientists have known about this material for decades, and that biomass and activity decrease rapidly with depth. With depth, larger organisms become less numerous, leaving bacteria and meiofauna to dominate abyssal life. But sparse abundance and biomass do not mean less diversity than in productive rainforests and coral reefs. This paradox has focused researchers on food supply to the deep sea and how it affects diversity.

When the Census deep-sea projects were developed less than a decade ago, almost 50 years had passed since Howard Sanders and Robert Hessler had killed the myth of a lifeless deep sea. They proposed that continental slope and abyssal plain sediments might rival tropical rainforests in diversity. Fred Grassle and Nancy Maciolek's hotly debated projection of 10 million deep-sea invertebrate species provoked new interest in the early 1990s. Nevertheless, most scientists accept that many species live in the deep, despite its apparent uniformity that sharply contrasts the three-dimensional structure of species-rich coral reefs.

The deep Southern Ocean has yielded a wealth of new species, from copepod and isopod crustaceans, corals and anemones to glass sponges. Some species of foraminiferans are distributed over 17,000 kilometers from pole to pole, while others change within only 1,000 kilometers. Even when focusing on better-known groups, MAR-ECO found about 10% of the species they sampled along the Mid-Atlantic Ridge were new, including fishes, multiple squids, sea stars, and sea cucumbers.

The Census used new tools (see Chapter 4) and coordinated deep-sea research around the world to show that varied habitats add to deep-sea biodiversity, building on discoveries in the 1980s and 1990s. The Census explored biodiversity in oxygen minimum zones or OMZs (where bacterial *respiration*, or breakdown of organic matter, depletes oxygen), seeps, whale carcasses, and dense deep-water coral beds with abundant fishes. They visited seamounts and manganese nodule fields on the abyssal seafloor, novel habitats of seabed brine pools and methane pock marks, and even mud volcanoes (Figure 8.2). As they studied the most poorly sampled habitat on Earth, CeDAMar standardized sampling tools and methods. With considerable data already in hand, COMARGE worked on the slopes to standardize new sampling, train taxonomists, and explore rich deep-sea diversity.

How do cold, pressure, and scarce food support a rich array of life? Three patterns of diversity known for decades complicate the answer. Diversity decreases away from the equator, biodiversity peaks at 2,000–3,000 meter mid-slope depths depending on taxa, and zonation of multiple species changes at key depths rather than independently of one another. Solving the mystery of diversity in the deep will inform understanding of patterns elsewhere on Earth, and guide ecologists and conservationists in identifying and protecting hot spots, as on land.

Patterns related to latitude and depth

Like elsewhere in the ocean and on land, deep-sea species are more diverse in the tropics and less diverse at the poles. The abundant foraminiferans in deep-sea sediments, isopod crustaceans, bivalves and gastropods, and fishes confirm this pattern. Along the Mid-Atlantic Ridge, far from the continental margins, the diversity of zooplankton to cephalopods and demersal fishes increases from cold Iceland toward the warm Azores. These patterns are clearer in the northern than the lesser-known southern hemisphere, where the pattern is questioned. Exceptions occur, as with deep-sea nematode threadworms that appear more diverse in productive temperate environments than tropical areas, and the debate continues.

Latitude itself does not drive these patterns, but is a proxy for energy, food, evolutionary and ecological history, and food supply that do. Temperature and salinity that create pattern in near-surface and coastal habitats (Chapters 5–7) are unlikely players because they vary little in the deep ocean, so studies have focused on productivity and basin age. In the abyss from the equator to the Southern Ocean, ongoing CeDAMar analysis will clarify whether there is a latitudinal gradient in the southern hemisphere.

Diversity generally peaks at 2,000–3,000 meters, as documented 30 years ago. However, the diversity of polychaete worms peaks at different depths in the eastern North Atlantic than their polychaete cousins in the tropical Atlantic, or compared with isopod crustaceans in the Southern Ocean. Neither megafauna on the Antarctic slope nor Mediterranean invertebrate groups peak at intermediate slope depths. Nonetheless, the pattern is reported widely. Why is this peak commonly found? Intermediate levels of available food may optimize diversity at mid-slope depths, where abundant or sparse food depresses diversity in shallower and deeper environments, respectively. Nonetheless, several studies find nematode diversity increases with surface phytoplankton production, and the pattern in mollusks and crustaceans are just the opposite, with no intermediate peak in either group.

To understand these on-again, off-again peaks, COMARGE analyzed 16 datasets from around the world from depths of 1,000 to 4,000 meters. Aside from some Southern Ocean taxa, they confirmed the mid-slope peak in most regions and taxa. Many exceptions were caused by oxygen minimum zones that intersect slopes at intermediate depths and create low-oxygen zones, where only microbes flourish. A rich terrestrial literature suggests too little food depresses diversity, as does too much, though some studies show a linear rather than hump-shaped relationship. Satellites and models have mapped surface phytoplankton, a proxy for food falling to the seafloor below that augments sparse measurements of deep-sea food supply. Bottom currents and topography confound these models. COMARGE found an intermediate food supply corresponded with peak diversity over depth. Their generalization explained only a small portion of diversity, in part because diversity was particularly variable, and thus unpredictable, where food input was low.

At ocean scales, faunal patterns in the abyss mirror food production far above and define biogeographic regions. Within weeks after a seasonal phytoplankton bloom at the surface, fauna thousands of meters below respond to the feast of falling *phytodetritus*, or decomposing phytoplankton, jellyfish falls, and whale carcasses. Responding to decadal meteorological changes like El Niño, food production on the surface changes, and diversity, distribution, and abundance of abyssal communities change accordingly. In the abyssal Indian Ocean, protistan foraminiferans were more diverse in more productive areas. In the Mediterranean abyss, more and better food increased abundance and diversity. In the abyssal central Pacific, more food increased polychaete diversity weakly. Perhaps faunal diversity would peak at intermediate supply in the abyss, but too little food reaches the seafloor to see the full range of response.

Bottom-temperature data from the *World Ocean Atlas* was the best predictor of deep-sea diversity, and adds to a long-known pattern of higher diversities in warmer climates, where increased metabolism promotes rapid speciation and thus diversity. Because the diversity of polychaete worms peaked at a boundary between water masses in the northeast Atlantic, gradients or *ecotones* between adjacent environments may be hot spots of deep-sea diversity.

COMARGE compared well-known zonation patterns and found that except for consistent zones along the junctions of continental shelves with upper slopes, depth zones varied among six regions, perhaps because of real geographic differences or perhaps because of sampling differences. But depth itself does not appear to explain the regional variation, particularly in light of depth differences in zones for different taxa within a region. Because the warm surface waters of the Mediterranean Sea penetrate to greater depths than elsewhere without changing the shelf-to-slope junction noted above, temperature does not explain zones. Similarly, the cold link between polar shelf and slope waters fails to explain the zonal patterns. Because the environment becomes more uniform with depth, zones span greater depth in deeper water. Nevertheless, fauna change sharply at 400–500 meters and at 1,500–2,500 meters.

In the Gulf of Mexico, food affects depth zones, but on the Mid-Atlantic Ridge, zonation was absent. The scarce food on and near the Mid-Atlantic Ridge may link pelagic and demersal fishes more closely than those on continental margins, and eliminate seafloor zonation. CenSeam found species and coral assemblages on seamounts were strongly linked with depth, but seamount zonation studies remain an opportunity for research.

Taxonomic consistency raises a great impediment to testing broad patterns in the deep ocean. Because the deep sea is so species rich and many species are rare, problems arise merging data from multiple studies and laboratories into a single analysis. COMARGE has melded comparable data for squat lobsters and nematodes, and others are underway. Comparable datasets, sometimes requiring genetic barcoding, have revealed that many deep-sea polychaete worms, bivalves, and foraminiferans once thought to be cosmopolitan are actually mixtures of look-alike species that differ among regions. Comparisons have found look-alike, but distinct abyssal nematodes and isopod crustaceans. With common databases and taxonomic workshops, CeDAMar can now compare broadly. They proposed that on abyssal plains under the Southern Ocean, larger isopods and polychaetes may be confined to productive areas, but smaller protists may be broadly distributed.

This story of latitude, depth, and accompanying zones cries for resolution of its uncertainty, and points to the isolated vents, seeps, and plains that offer contrasts like those between gardens and deserts, and perhaps insight from local scales to the broader ocean.

When are "deserts" not deserts?

Planet Ocean offers many specialized environments for species (Figure 8.3), some that are toxic for many, but offer opportunities for others. On continental slopes, the environment changes over distances of centimeters to thousands of kilometers and timescales of millions of years to seasons. These changes offer diverse niches for many species, including some without relatives on land. Abundances and diversity of organisms are less in the deep open water down to the seafloor because food particles sink quickly through and accumulate to fuel the seafloor community. Unfortunately there are fewer data in this mid-range than any other ocean environment (Figure 8.4). Despite forbidding depth, darkness, pressure, heat, and chemistry, specialized forms have evolved in hydrothermal vents and seeps.

Vent specialists recognize six biological provinces defined by geological history and ocean circulation. Where the seafloor spreads or one tectonic plate slips below another (Figure 8.5), hydrothermal vents open up. There, seawater percolates through cracks into the upper mantle, is superheated at high pressure, and enriched in hydrogen sulfide, methane, and minerals. These provinces have evolved their own suite of species, some endemic and some cryptic.

Figure 8.3 Deep-sea habitats.
Habitats in the deep sea span productive vents and seeps, complex cobble and seamounts, and rolling plains of sediment.
(a) A deep-water coral habitat on the upper continental slope of Nova Scotia, Canada. Sampling this region of the Discovery Corridor was part of the Canadian Healthy Oceans Network (CHONe) that grew from the Census. The Network explores deep-water habitats from coral areas like this one, as well as exploring life in deep-sea sediments.

(d)

(e)

(f)

(b) Seeping natural gas from the seafloor fuels a microbial reef 250 meters beneath the northwest Black Sea. This community is fed by specialized microbes that use natural gas rather than photosynthesis, which fuels most biological communities.

(c) Census scientists sailed on the Australian icebreaker *Aurora Australis* to explore diverse deep-water coral near Antarctica, 725 meters below the surface.

(d) Around cold seeps, tubeworms may live singly or aggregate in fields of thousands, exploiting methane and sulfides bubbling from the seafloor. Some of the species gather around hydrothermal vents, too, but many live only near the cold seeps.

(e) In deep water near the Larsen B area of Antarctica, abundant sea cucumbers all turn to feed in the same direction. Relatives of sea stars, sea cucumbers help cycle the carbon in food through deep-sea ecosystems.

(f) An aggregation of brittlestars on the Macquarie Ridge off New Zealand extend their arms to catch food in passing currents.

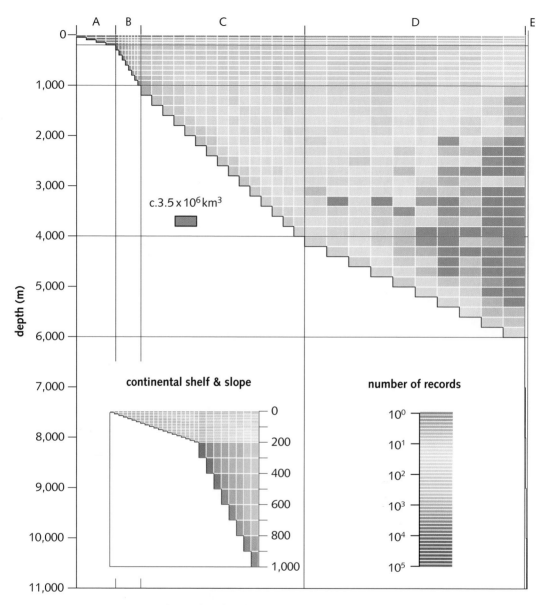

Figure 8.4 The explored and unexplored ocean.

The number of OBIS samples in five zones of the ocean, labeled A–E, shows the dearth of samples between 2,000 and 6,000 meters deep in the water and on the deep seafloor. The horizontal axis splits the ocean into depth zones (A = 0–200 meters, B = 200–1,000 meters, C = 1,000–4,000 meters, D = 4,000–6,000 meters and E = >6,000 meters), where the width of each zone is proportional to global surface area. The number of data records in each cell is standardized to the volume of water in that cell and $\log_{10}$-transformed. The inset shows in greater detail the continental shelf and slope, where the most records are found.

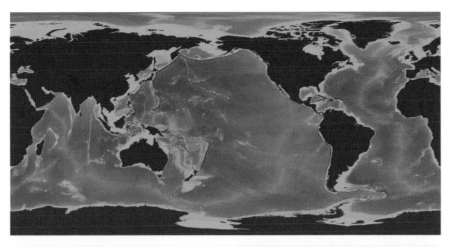

Figure 8.5 Long ocean ridges divide vast abyssal plains on the seafloor.
The distinct mid-ocean ridges on the seafloor demarcate some edges of the plates
that cover the Earth. During geological time, the slow-moving plates rearrange
the seafloor. The enlarged bottom panel shows the North Atlantic and its Mid-
Atlantic Ridge.

Vents remain to be discovered in unexplored regions of the deep where the seafloor is spreading and slipping.

ChEss discovered vents in the Lau Basin in the southwest Pacific, in the East Pacific Rise in the southeast Pacific, and on the southwest Indian Ridge of the southern Indian Ocean. These new locations built on known biological provinces, but new vent fields in the South Atlantic portion of the Mid-Atlantic Ridge may change current provinces. ChEss also discovered the hottest (407 °C) and deepest (5,000 meters) vents known so far.

Where the seafloor spreads ultra-slowly near the Azores, temperature and venting is low and the faunas of newly discovered vents resemble slope and seamount communities rather than other vents. ChEss discovered more than 10 new seep sites north of New Zealand, including one 135,000-square-meter area that is one of the largest globally. Vent and seep communities overlapping other areas of the western Pacific may represent a new province of seep and vent faunas. In the Nile Deep Sea Fan, ChEss also found new hydrocarbon seeps at 1,000–3,500 meters depth, and hydrothermal activity in the Norwegian Sea and Arctic Ocean.

Across the deep-sea floor and through evolutionary time, faunas have colonized and diversified. Shallow-water isopods colonized the deep ocean during four periods, diversifying in abyssal habitats. Evolutionary history links shelf, slope, and abyssal environments. In the Southern Ocean, species of isopods and polychaetes invaded slopes from shelf or abyssal origins, whereas shelf and slope mollusks differ. These species pools comprise a diverse Southern Ocean fauna that surprised taxonomic experts. In contrast, isolation of the Mediterranean Sea from the Atlantic by the shallow sill at the Strait of Gibraltar, in concert with major disturbance, has created an impoverished deep-sea biota that lacks major groups of echinoderms and glass sponges. Still, Mediterranean nematodes are relatively rich and areas that trap food material create hot spots of activity.

Generally, abyssal species appear widely distributed among basins, some more so than on slope depths. Most, but not all polychaete worms occur widely. Some cosmopolitan polychaetes, crustaceans, and foraminiferans span basins, while others are limited to specific regions. Many abyssal species colonized from the continental slope and their widespread distributions reflect slow adaptation and evolution in the abyss. Nevertheless, areas of the abyssal Pacific have evolved characteristic fauna and many foraminiferans and nematodes are truly abyssal in origin, distinct from slopes and distinct from other regions.

Over hundreds to thousands of kilometers, ocean currents move water masses on and off continental slopes, changing oxygen concentrations, temperatures, and currents. These, in turn, alter patterns of large seafloor invertebrates and fishes. Currents create patterns of sediment and species across hundreds of kilometers. Sediments that cover slope environments are less uniform than they look. Steady winds push surface waters offshore and nutrient-rich waters rise to the surface. Biological production increases and more detritus sinks to the seafloor. Beneath the Canary, California, and Humboldt Currents, abundant detritus decays, depleting oxygen and generating hydrogen sulfide that supports mats of specialized bacteria that dominate oxygen minimum zones.

Some canyons and ridges span hundreds of kilometers, while mid-ocean ridges extend thousands of kilometers. Canyons channel food from continental shelves downslope, supporting abundant fish, sponges, and other invertebrates. Sediment composition and elevated food material often contrast with the surrounding slope, elevating biomass and abundance, but depressing diversity. Canyons link slopes and the abyss to adjacent shelf habitats. On the continental slope of the Gulf of Guinea off Africa, the Congo River Canyon and Channel transport food for megafauna and others into deeper water on the abyssal plain.

Seamounts also contrast with surrounding abyssal plains and continental slopes. Seamounts differ in depth below the surface and height above the seabed, isolation, and age. Flat-topped seamounts, *guyots*, accumulate sediments, but steep sides expose bedrock and coarser sediments. Currents, oxygen and chemistry, light, temperature, substratum, and productivity also influence seamount biota. Only about 250 have been sampled enough to say much about their species or to draw broad conclusions on diversity and endemism.

Productive seamounts have been fished since the fourteenth century, and attracted industrial trawling in the 1970s. Their unique isolation, geology, and productivity have intrigued biologists since the late 1950s and have recently generated a pulse of interest catalyzed by CenSeam. The CenSeam SeamountsOnline database coordinated research, enabled broad-scale analyses, and revealed key knowledge gaps.

How do seamounts contribute to local and regional species pools? Though historically considered diversity hot spots, that view is changing. Some species on the most isolated seamounts are endemic, but recent exploration finds the same species live on seamounts and adjacent deep-sea areas, and there is no evidence of elevated seamount diversity. Many seamount corals and fishes occur widely and sometimes globally, genetically similar to populations on margins or near oceanic islands. But differing proportions of species may be one way that seamounts contribute to the riot of species in the deep ocean.

Seamounts may contribute more to deep-sea abundance than to diversity. Upwelling around seamounts recirculates water, amplifies waves, and creates internal waves that enhance productivity and retain offspring. Phytoplankton near seamounts are abundant, but insufficient for the many fishes and other organisms that must also depend on food carried in currents that accelerate around seamounts, or on suitable prey items in the migrating zooplankton. Scientists have found only mixed evidence that the complex three-dimensional seamount corals increase diversity of other species, although cruises along the Mid-Atlantic Ridge found 40 species of deep-water corals with elevated fish diversity.

The Mid-Atlantic Ridge (MAR) bisects the Atlantic Ocean as it extends above the abyssal plain (Figure 8.5). Spanning water masses and latitude, it offers unique opportunities for discovery on diversity, distribution, and abundance. MAR-ECO found abundant copepods in the upper 100 meters, euphausiid peaks in the upper 200 meters, and decapod peaks at 200–700 meters. Gelatinous species peaked from 400–900 meters. In contrast to sharp declines below 1,000 meters in most ecosystems, open-sea *pelagic* fish biomass peaked at 1,500–2,300 meters, perhaps because the benthos of the ridge provided a food source for the pelagic fishes. Abundant *demersal*, or bottom, fishes at the summit of the ridge declined with depth. For open-sea and bottom fishes, depth was more important than latitude. Surprisingly, sharks, skates, and rays are rarely, perhaps never, found below 3,000 meters, where their energy demands and the limited food in the abyss may exclude them.

Along the Mid-Atlantic Ridge, water masses define zooplankton pattern. The *Sub-Polar Front* between water masses creates a barrier for zooplankton, seabirds, and whales. Because bottom topography shapes water masses, it influences dolphins, sei whales, and sperm whales that congregate in surface waters thousands of meters above. Closer to the seafloor, large, adult fishes dominate, many in spawning or near-spawning condition. Ridge systems may be spawning and biomass hot spots, a hypothesis strengthened by the discovery of small, rare juvenile orange roughy and postlarval blue hake.

Manganese nodules on the abyssal plains create benthic habitat on scales of centimeters to kilometers, with higher species diversity than other areas nearby. During a 20-year study of the Porcupine Abyssal Plain southwest of Ireland, CeDAMar researchers discovered major increases in the lowly sea cucumber *Amperima rosea*, a major player in deep-ocean carbon cycling. Some species did not change, but protists and meiofauna were affected by increased food and *Amperima* burrowing. While small bacteria and meiofauna dominate numbers on abyssal plains, larger species dominate energy transfer. These changes were linked to increased food from surface waters as conditions changed. Though the 20-year study focused on changes over time, it also illustrates how food descending from above affects species composition far below.

Where abundant food accumulates and decomposes, methane and gas hydrates bubble up and support specialized bacteria on wood, whale carcasses, and OMZs. Communities on whale carcasses and wood falls, pockmarks in sediments, mud volcanoes, large microbial mats, and methane and hydrocarbon seeps all depend on these specialized bacteria and archaea. These habitats only span the area of a few football fields or less. Animals have ranges of tolerance to low oxygen and methane concentrations that define where they live. At smaller scales still, animals clustered at seeps modify chemistry to create unique habitats, providing habitat, refuges, and food. On smaller scales of centimeters to millimeters, small invertebrates create habitats for microbes, adding to the species pool. In addition to habitat, biological succession and inhibition contribute to species patterns and abundances.

Despite their heat and toxic emissions, hydrothermal vents support much life. The water cools from 350 °C to the ambient 2–3 °C in just a few meters. Until diluted, the abundant hydrogen sulfide, dissolved metals, and absence of oxygen start a gradient lethal to most life. These gradients create a range of microenvironments for free-living microbes and microbes living inside larger hydrothermal vent organisms. Clams and tubeworms at vents have genes, enzymes, and other adaptations to tolerate these conditions and feed from the intense microbial production. Metals in vent fluids accumulate in animals and pass from prey to predators. As in seeps, vent animals create microhabitats for other species between their tubes and shells, adding to the diversity of vent species that range from abundant to rare. Vent isolation encourages endemism, leaving 70% of known vent species unique to a region. Vent larvae are sometimes retained within the vents they came from.

These small-scale environments are transient in the staid deep sea. As minerals flow from vents they precipitate to form solid chimneys that grow to 40 meters and eventually topple from instability. Vents ebb and flow as new cracks form in the crust and others close. The landscape of the habitats may persist for decades, years, or less, supporting a few specialized species, but then disappear. The desert oases come and go.

Stepping stones to a diverse deep sea

Many species occur only in the isolated vents or seeps. Some live at hydrothermal vents, but not elsewhere, whereas others also live on whale carcasses. Because most of these species move only as larvae, how do they find these "islands" of heat and toxins, and evolve endemic only to that location? However the inhabitants find them, the isolated vents and seeps add deep-sea diversity. Whale carcasses support a rich fauna locally, including a newly discovered species of lancelet and rare species of a bottom-living comb jelly and snail. A single whale carcass has as many species as global seeps combined or all species from a geographic cluster of vents. Unique species in these specialized environments are modest in number, but add significantly to regional diversity.

   Habitats of many types and sizes (Figure 8.3) transform the "desert" landscape under the ocean into complex environments defined by temperature, seafloor composition, necessary food, and oxygen and chemical gradients. Biology also creates habitat. Microbes feed their living hosts in a *mutualism*, where one organism depends on another. Some species wait until others first make the habitat suitable, so crabs may not appear at vents until tubeworms colonize first.

The deep human footprint

It hardly seems possible that human activities on land leak out to the remote and immense deep ocean, but they do. They act in concert with activities at sea to create an array of current and future threats. Recent evidence links biodiversity to deep ocean health, where seafloor biodiversity enhances productivity and cycling of material. This COMARGE discovery suggests how the loss of species could compromise health in the deep sea.

Pollutants from land make their way via drainage into the nearshore environment and then into the deep ocean. They flow through canyons that carry lead and other metal contaminants. Elevated concentrations of toxic DDT and PCBs show up in deep-sea fauna.

Only decades after depletions of coastal fisheries drove fishermen deeper and deeper, some deep-sea species are now depleted and endangered. Lucrative deep-sea fisheries collapsed and simultaneously drove non-target species to near extinction as fishing gear destroyed deep-water corals. Bottom trawling down to 1,600 meters may decrease abundance to 2,500 meters as fish move into vacated habitat. The footprint of deep-sea fishing is broader than the targeted areas or species as trawls bring up tons of damaged biota that is unceremoniously dumped over the side.

In addition to uprooting fragile sessile fauna on seamounts, fishing removes fauna that are slow to reproduce, produce only a few offspring, grow slowly, and live long. Bottom trawls quickly eliminate deep-water coral habitat and other invertebrates. Fishermen quickly find fish populations concentrated around seamounts and exploit them, bringing rapid collapse. Most deep-sea fisheries are biologically unsustainable, perhaps a moot point because management and enforcement are difficult in international waters where many seamounts occur. Fishing has affected even remote and patchy seep environments on the California margin, the Chilean margin, and newly discovered seeps near New Zealand, where fishing gear arrived before scientists. Recovery from fishing takes decades or longer for seamount corals and fishes.

Cost has prohibited mining the deep-sea bed, but rising metal prices may make deep-sea mining profitable by the year 2025. Millions of square-kilometers of the abyssal seafloor covered with golf-ball to softball-sized manganese nodules rich in iron, manganese, copper, nickel, and cobalt sit deep below low productivity waters of the Pacific and Indian Oceans. Foreseeing nodule mining, nations have staked seabed claims. To test the vulnerability of the Pacific abyss, CeDAMar investigated whether most manganese nodule species are endemic and vulnerable or cosmopolitan with refuges elsewhere. Some suggest abyssal species reproduce on the continental slope and marginally survive in the abyss. CeDAMar, however, found a characteristic fauna in the Pacific abyss, with many foraminiferans and nematodes unique to the abyss. Molecular analysis uncovered many cryptic polychaete species endemic to that area.

Mining would disturb seafloor, seamount, and hydrothermal vent environments. Companies now plan to mine the gold, silver, copper, and zinc in chimney-like deposits at vents. They have already begun environmental-impact studies with plans to exploit these deposits within the next decade, though knowing little of the consequences for vents. Seamounts are also hot spots for cobalt-rich manganese crusts, manganese nodules, and polymetallic sulfides that may be exploited for precious and base metals.

Oil drilling pumps hydrocarbons from thousands of meters below the ocean's surface. Drill-cutting spoils smother organisms and add enduring toxic chemicals with poorly known effects out to tens of kilometers. Other commercial activities threaten, from mining methane hydrates to storing CO_2 in seabeds or fixing it at the surface by iron fertilization. Knowledge of CO_2 disposal effects on benthic organisms is equivocal. Finally, scientists themselves must not damage or spoil fragile hydrothermal vents and seeps as they work to uncover their secrets.

Global warming could affect the deep ocean. Global warming might stratify tropical and temperate oceans, reducing nutrients for phytoplankton and reducing food export into the deep. Lower abyssal abundance and changing species will follow, and because changes are expected over massive areas, extinctions are likely. Less mixing of oxygen into the deep would widen OMZs in the tropics. The northward extension of Humboldt squid in warming waters of the Oregon, Washington, and Alaska coasts may be harbingers of changes that warming will cause.

The global ocean paid a price for taking up almost a third of the additional CO_2 that humans added to the atmosphere during the twentieth century. Dissolved CO_2 increased ocean acidity measurably and acidity dissolves calcium carbonate, the stuff of coral skeletons, among others. In the North Pacific, coral distributions are already changing. Echinoderms, mollusks, foraminiferans, and corals are candidates to suffer this century if acidification worsens, with cascading effects on other species. Low oxygen waters in OMZs are already more acidic than elsewhere and as they expand will amplify acidification effects.

The immensity, diversity, and unknown mysteries of the deep ocean heighten fears we will harm it. Its immense diversity and hot spots of endemism suggests a trove of "sunken treasure" that trumps any pirate's chest. One deep-sea trawl in the deep Mediterranean retrieved more trash than

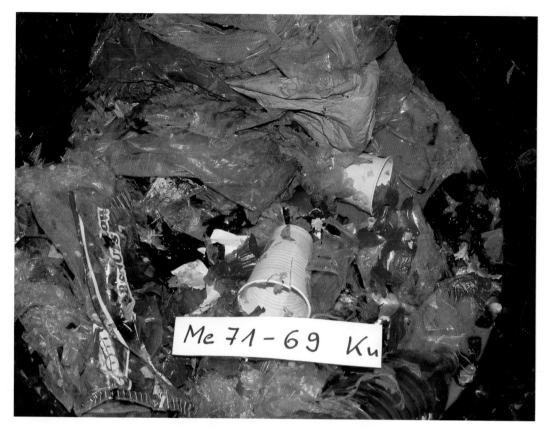

Figure 8.6 More garbage than life on the seafloor.
Searching for marine life, Census researchers towed a trawl at 3,000 meters beneath
the Eastern Mediterranean near Crete. Instead of living creatures, they hauled up a net
filled with trash. Passing ships and currents had carried and deposited the trash,
which decayed little on the deep seafloor.

animals (Figure 8.6). As a 20-year-old undergraduate, I dissected a redfish
hardly big enough for a single meal. As I worked, a thought gave me
pause: the redfish under my knife was much older than I was.

But the deep-sea story is mostly a very good story and an exciting one.
Scientists aboard the *Challenger* expedition could hardly have dreamed of
the array of tools available today (Figure 8.7) that opens up the expanse of
diversity and unknown mysteries of the pristine deep to scientific exploration.
By playing their peculiar role of discovering causes and effects well, scientists
can convert unknowns into knowns and lessen the feared harms to the
future ocean.

Figure 8.7 Sampling in the deep.
Census scientists lowered samplers, dove down in submersibles, or launched autonomous vehicles to explore the deep. **(a)** Wood Hole Oceanographic Institution's Autonomous Benthic Explorer (ABE). ABE was lost off the coast of Chile in March 2010 after 16 years of service exploring the global ocean. **(b)** Submersible *Nautile* carried French explorers down to study life on the seafloor around hot and toxic hydrothermal vents.

(c) Acoustic multibeam sonar map of a seamount atop the Macquarie Ridge. The ridge extends 1,400 kilometers southwest of New Zealand into the Southern Ocean. The image, magnified fivefold in the vertical dimension, will guide future sampling.

(d) Near a vent 3 kilometers beneath the equatorial Atlantic, Census researchers used the remotely operated vehicle QUEST to find shrimp and other marine life.

(e) A multicorer is lowered over the rail of CGS *Hudson* to the seafloor thousands of meters below to collect sediments and their fauna with minimal disturbance.

BIBLIOGRAPHY

Aboim, M. A., G. M. Menezes, T. Schlitt, and A. D. Rogers, Genetic structure and history of populations of the deep-sea fish *Helicolenus dactylopterus* (Delaroche, 1809) inferred from mtDNA sequence analysis. *Mol. Ecol.*, **14**:5 (2005), 1343–54.

Adams, D. K. and L. S. Mullineaux, Supply of gastropod larvae to hydrothermal vents reflects transport from local larval sources. *Limnol. Oceanogr.*, **53**:5 (2008), 1945–55.

Allen, A. P., J. H. Brown, and J. F. Gillooly, Global biodiversity, biochemical kinetics, and the energetic-equivalence rule. *Science*, **297**:5586 (2002), 1545–8.

Althaus, F., A. Williams, T. Schlacher, *et al.*, Impacts of bottom trawling on deep-coral ecosystems of seamounts are long-lasting. *Mar. Ecol. Prog. Ser.*, **397** (2009), 279–94.

Amaral-Zettler, L., L. F. Artigas, J. Baross, *et al.*, A global census of marine microbes. In A. D. McIntyre (ed.), *Life in the World's Oceans: Diversity, Distribution, and Abundance* (Oxford: Blackwell Publishing Ltd., 2010) pp. 223–45.

Arima, S. and K. Nagakura, Mercury and selenium content of odontoceti. *Bull. Jpn Soc. Sci. Fish.*, **45**:5 (1979), 623–6.

Baba, K., E. Macpherson, G. C. B. Poore, *et al.*, Catalogue of squat lobsters of the world (Crustacea: Decapoda: Anomura – families Chirostylidae, Galatheidae and Kiwaidae). *Zootaxa*, **1905** (2008), 3–220.

Baco, A. R., Exploration for deep-sea corals on North Pacific seamounts and islands. *Oceanography*, **20**:4 (2007), 108–17.

Baco, A. R., A. A. Rowden, L. A. Levin, C. R. Smith, and D. A. Bowden, Initial characterization of cold seep faunal communities on the New Zealand Hikurangi margin. *Mar. Geol.* **272** (2010), 251–9.

Baco, A. R. and C. R. Smith, High species richness in deep-sea chemoautotrophic whale skeleton communities. *Mar. Ecol. Prog. Ser.*, **260** (2003), 109–14.

Baguley, J. G., P. A. Montagna, L. J. Hyde, and G. T. Rowe, Metazoan meiofauna biomass, grazing, and weight-dependent respiration in the Northern Gulf of Mexico deep sea. *Deep Sea Res. II*, **55**:24–26 (2008), 2607–16.

Baker, M. C. and C. R. German, Going for gold! Who will win in the race to exploit ores from the deep sea? *Ocean Chall.*, **16** (2009), 10–17.

Baker, M. C., E. Z. Ramirez-Llodra, P. A. Tyler, *et al.*, Biogeography, ecology and vulnerability of chemosynthetic ecosystems in the deep sea. In A. D. McIntyre (ed.), *Life in the World's Oceans: Diversity, Distribution, and Abundance* (Oxford: Blackwell Publishing Ltd., 2010), pp. 161–182.

Behrenfeld, M. J. and P. G. Falkowski, A consumer's guide to phytoplankton primary productivity models. *Limnol. Oceanogr.*, **42**:7 (1997), 1479–91.

Berger, W. H., K. Fischer, C. Lai, and G. Wu, *Ocean Productivity and Organic Carbon Flux* (San Diego: University of California, 1987).

Bergquist, D. C., J. T. Eckner, I. A. Urcuyo, *et al.*, Using stable isotopes and quantitative community characteristics to determine a local hydrothermal vent food web. *Mar. Ecol. Prog. Ser.*, **330** (2007), 49–65.

Bergquist, D. C., C. Fleckenstein, J. Knisel, *et al.*, Variations in seep mussel bed communities along physical and chemical environmental gradients. *Mar. Ecol. Prog. Ser.*, **293** (2005), 99–108.

Bergstad, O. A., G. Menezes, and A. S. Hoines, Demersal fish on a mid-ocean ridge: distribution patterns and structuring factors. *Deep Sea Res. II*, **55**:1–2 (2008), 185–202.

Bernhard, J. M., J. P. Barry, K. R. Buck, and V. R. Starczak, Impact of intentionally injected carbon dioxide hydrate on deep-sea benthic foraminiferal survival. *Global Change Biol.*, **15**:8 (2009), 2078–88.

Bertics, V. J. and W. Ziebis, Biodiversity of benthic microbial communities in bioturbated coastal sediments is controlled by geochemical microniches. *ISME J.*, **3**:11 (2009), 1269–85.

Bett, B. J., M. G. Malzone, B. E. Narayanaswamy, and B. D. Wigham, Temporal variability in phytodetritus and megabenthic activity at the seabed in the deep northeast Atlantic. *Prog. Oceanogr.*, **50**:1–4 (2001), 349–68.

Bettencourt, R., P. Dando, P. Collins, *et al.*, Innate immunity in the deep sea hydrothermal vent mussel *Bathymodiolus azoricus*. *Comp. Biochem. Physiol. A, Mol. Integr. Physiol.*, **152**:2 (2009), 278–89.

Bettencourt, R., P. Roch, S. Stefanni, *et al.*, Deep sea immunity: unveiling immune constituents from the hydrothermal vent mussel *Bathymodiolus azoricus*. *Mar. Environ. Res.*, **64**:2 (2007), 108–27.

Billett, D. S. M., B. J. Bett, C. L. Jacobs, I. P. Rouse, and B. D. Wigham, Mass deposition of jellyfish in the deep Arabian Sea. *Limnol. Oceanogr.*, **51**:5 (2006), 2077–83.

Billett, D. S. M., B. J. Bett, W. D. K. Reid, B. Boorman, and I. G. Priede, Long-term change in the abyssal NE Atlantic: the "*Amperima* Event" revisited. *Deep Sea Res. II* (2010), 1406–17.

Billett, D. S. M., B. J. Bett, A. L. Rice, *et al.*, Long-term change in the megabenthos of the Porcupine Abyssal Plain (NE Atlantic). *Prog. Oceanogr.*, **50**:1–4 (2001), 325–48.

Billett, D. S. M., R. S. Lampitt, A. L. Rice, and R. F. C. Mantoura, Seasonal sedimentation of phytoplankton to the deep-sea benthos. *Nature*, **302**:5908 (1983), 520–2.

Bouchet, P. and A. Waren, Planktotrophic larval development in deep-water gastropods. *Sarsia*, **64** (1979), 37–40.

Brand, G. L., R. V. Horak, N. Le Bris, *et al.*, Hypotaurine and thiotaurine as indicators of sulfide exposure in bivalves and vestimentiferans from hydrothermal vents and cold seeps. *Mar. Ecol.*, **28**:1 (2007), 208–18.

Brandt, A., C. De Broyer, I. De Mesel, *et al.*, The biodiversity of the deep Southern Ocean benthos. *Philos. Trans. R. Soc. Lond. B Biol. Sci.*, **362**:1477 (2007), 39–66.

Brandt, A., K. Ellingsen, S. Brix, W. Brokeland, and M. Malyutina, Southern Ocean deep-sea isopod species richness (Crustacea, Malacostraca): influences of depth, latitude and longitude. *Polar Biol.*, **28**:4 (2005), 284–9.

Brandt, A., K. Linse, and M. Schuller, Bathymetric distribution patterns of Southern Ocean macrofaunal taxa: Bivalvia, Gastropoda, Isopoda and Polychaeta. *Deep Sea Res. I*, **56**:11 (2009), 2013–25.

Brewin, P. E., K. I. Stocks, and G. Menezes, A history of seamount research. In T. Pitcher, T. Morato, P. J. B. Hart, M. R. Clark, N. Haggan, and R. S. Santos (eds.), *Seamounts: Ecology, Fisheries and Conservation* (Oxford: Wiley-Blackwell, 2007), pp. 41–61.

Briggs, J. C., Species diversity: land and sea compared. *Syst. Biol.*, **43**:1 (1994), 130–5.

Brown, J. H., J. F. Gillooly, A. P. Allen, V. M. Savage, and G. B. West, Toward a metabolic theory of ecology. *Ecology*, **85**:7 (2004), 1771–89.

Cardigos, F., A. Colaco, P. R. Dando, *et al.*, Shallow water hydrothermal vent field fluids and communities of the D. Joao de Castro Seamount (Azores). *Chem. Geol.*, **224**:1–3 (2005), 153–68.

Carney, R. S., Zonation of deep biota on continental margins. *Oceanogr. Mar. Biol. Annu. Rev.*, **43** (2005), 211–78.

Carney, R. S., R. L. Haedrich, and G. T. Rowe, Zonation of fauna in the deep sea. In G. T. Rowe (ed.), *The Sea, Vol. 8, Deep Sea Biology* (New York: Wiley, 1983), pp. 371–98.

Carney, S. L., J. F. Flores, K. M. Orobona, *et al.*, Environmental differences in hemoglobin gene expression in the hydrothermal vent tubeworm, *Ridgeia piscesae*. *Comp. Biochem. Physiol. B, Biochem. Mol. Biol.*, **146**:3 (2007), 326–37.

Casey, J. M. and R. A. Myers, Near extinction of a large, widely distributed fish. *Science*, **281**:5377 (1998), 690–2.

Chernova, N. V. and P. R. Moller, A new snailfish, *Paraliparis nigellus* sp nov (Scorpaeniformes, Liparidae), from the Northern Mid-Atlantic Ridge – with notes on occurrence of *Psednos* in the area. *Mar. Biol. Res.*, **4**:5 (2008), 369–75.

Clark, M., Fisheries for orange roughy (*Hoplostethus atlanticus*) on seamounts in New Zealand. *Oceanol. Acta*, *22*:6 (1999), 593–602.

Clark, M., Are deepwater fisheries sustainable? – the example of orange roughy (*Hoplostethus atlanticus*) in New Zealand. *Fish. Res.*, **51**:2–3 (2001), 123–35.

Clark, M. and R. L. O'Driscoll, Deepwater fisheries and aspects of their impact on seamount habitat in New Zealand. *J. NW Atlantic Fish. Sci.*, **31** (2003), 441–58.

Clark, M., A. A. Rowden, T. Schlacher, *et al.*, The ecology of seamounts: structure, function, and human impacts. *Annu. Rev. Mar. Sci.*, **2** (2010), 253–78.

Clark, M. R. and A. A. Rowden, Effect of deepwater trawling on the macro-invertebrate assemblages of seamounts on the Chatham Rise, New Zealand. *Deep Sea Res. I*, **56**:9 (2009), 1540–54.

Clark, M. R., V. I. Vinnichenko, J. D. M. Gordon, *et al.*, Large-scale distant-water trawl fisheries on seamounts. In T. Pitcher, T. Morato, P. J. B. Hart, M. R. Clark, N. Haggan, and R. S. Santos (eds.), *Seamounts: Ecology, Fisheries and Conservation* (Oxford: Wiley-Blackwell, 2007), pp. 361–99.

Colaco, A., P. Bustamante, Y. Fouquet, P. M. Sarradin, and R. Serrao-Santos, Bioaccumulation of Hg, Cu, and Zn in the Azores triple junction hydrothermal vent fields food web. *Chemosphere*, **65**:11 (2006), 2260–7.

Company, R., A. Serafim, R. Cosson, *et al.*, Temporal variation in the antioxidant defence system and lipid peroxidation in the gills and mantle of hydrothermal vent mussel *Bathymodiolus azoricus*. *Deep Sea Res. I*, **53**:7 (2006), 1101–16.

Connelly, D. P., C. R. German, M. Asada, *et al.*, Hydrothermal activity on the ultra-slow spreading southern Knipovich Ridge. *Geochem. Geophys. Geosyst.*, **8** (2007), Q08013 doi: 10.1029/2007GC001652.

Consalvey, M., M. R. Clark, A. A. Rowden, and K. I. Stocks, Life on seamounts. In A. D. McIntyre (ed.), *Life in the World's Oceans: Diversity, Distribution, and Abundance* (Oxford: Blackwell Publishing Ltd., 2010), pp. 123–38.

Cordes, E. E., D. C. Bergquist, and C. R. Fisher, Macroecology of Gulf of Mexico cold seeps. *Annu. Rev. Mar. Sci.*, **1** (2009), 143–68.

Cordes, E. E., S. L. Carney, S. Hourdez, *et al.*, Cold seeps of the deep Gulf of Mexico: community structure and biogeographic comparisons to Atlantic equatorial belt seep communities. *Deep Sea Res. I*, **54**:4 (2007), 637–53.

Cordes, E., M. R. Cunha, G. Galéron, *et al.*, The influence of geological, geochemical, and biogenic habitat heterogeneity on seep biodiversity. *Mar. Ecol.*, **31**:1 (2010), 54–65.

Cordes, E. E., S. Hourdez, B. L. Predmore, M. L. Redding, and C. R. Fisher, Succession of hydrocarbon seep communities associated with the long-lived foundation species *Lamellibrachia luymesi*. *Mar. Ecol. Prog. Ser.*, **305** (2005), 17–29.

Cosson-Sarradin, N., M. Sibuet, G. L. J. Paterson, and A. Vangriesheim, Polychaete diversity at tropical Atlantic deep-sea sites: environmental effects. *Mar. Ecol. Prog. Ser.*, **165** (1998), 173–85.

Culver, S. J. and M. A. Buzas, Global latitudinal species diversity gradient in deep-sea benthic foraminifera. *Deep Sea Res. I*, **47**:2 (2000), 259–75.

Currie, D. J., Energy and large-scale patterns of animal and plant species. *Am. Nat.*, **137** (1991), 27–41.

Currie, D. R. and L. R. Isaacs, Impact of exploratory offshore drilling on benthic communities in the Minerva gas field, Port Campbell, Australia. *Mar. Environ. Res.*, **59**:3 (2005), 217–33.

Danovaro, R., J. B. Company, C. Corinaldesi, *et al.*, Deep-sea biodiversity in the Mediteranean Sea: the known, the unknown, and the unknowable. *PLoS ONE* **5**:8 (2010), e11832, doi: 10.1371 journal.pone 0011832.

Danovaro, R., C. Gambi, A. Dell'Anno, *et al.*, Exponential decline of deep-sea ecosystem functioning linked to benthic biodiversity loss. *Curr. Biol.*, **18**:1 (2008), 1–8.

Dattagupta, S., M. A. Arthur, and C. R. Fisher, Modification of sediment geochemistry by the hydrocarbon seep tubeworm *Lamellibrachia luymesi*: a combined empirical and modeling approach. *Geochim. Cosmochim. Acta*, **72**:9 (2008), 2298–315.

Dattagupta, S., J. Martin, S. M. Liao, R. S. Carney, and C. R. Fisher, Deep-sea hydrocarbon seep gastropod *Bathynerita naticoidea* responds to cues from the habitat-providing mussel *Bathymodiolus childressi*. *Mar. Ecol.*, **28**:1 (2007), 193–8.

Davies, A. J., J. M. Roberts, and J. Hall-Spencer, Preserving deep-sea natural heritage: emerging issues in offshore conservation and management. *Biol. Conserv.*, **138**:3–4 (2007), 299–312.

Desbruyeres, D., N. Segonzac, and M. Bright, *Handbook of Deep-sea Hydrothermal Vent Fauna*, 2nd edn (Denisia: Linz, 2006).

Devey, C. W., C. R. Fisher, and S. Scott, Responsible science at hydrothermal vents. *Oceanography*, **20**:1 (2007), 162–71.

Devine, J. A., K. D. Baker, and R. L. Haedrich, Fisheries: deep-sea fishes qualify as endangered. *Nature*, **439**:7072 (2006), 29.

Dilman, A. B., Asteroid fauna of the northern Mid-Atlantic Ridge with description of a new species *Hymenasterides mironovi* sp nov. *Mar. Biol. Res.*, **4**:1–2 (2008), 131–51.

Doksaeter, L., E. Olsen, L. Nottestad, and A. Ferno, Distribution and feeding ecology of dolphins along the Mid-Atlantic Ridge between Iceland and the Azores. *Deep Sea Res. II*, **55**:1–2 (2008), 243–53.

Dower, J., H. Freeland, and K. Juniper, A strong biological response to oceanic flow past Cobb Seamount. *Deep Sea Res. I*, **39**:7–8 (1992), 1139–45.

Dubilier, N., C. Bergin, and C. Lott, Symbiotic diversity in marine animals: the art of harnessing chemosynthesis. *Nat. Rev. Microbiol.*, **6**:10 (2008), 725–40.

Duperron, S., C. Bergin, F. Zielinski, *et al.*, A dual symbiosis shared by two mussel species, *Bathymodiolus azoricus* and *Bathymodiolus puteoserpentis* (Bivalvia: Mytilidae), from hydrothermal vents along the northern Mid-Atlantic Ridge. *Environ. Microbiol.*, **8**:8 (2006), 1441–7.

Duperron, S., S. Halary, J. Lorion, M. Sibuet, and F. Gaill, Unexpected co-occurrence of six bacterial symbionts in the gills of the cold seep mussel *Idas sp* (Bivalvia: Mytilidae). *Environ. Microbiol.*, **10**:2 (2008), 433–45.

Duperron, S., M. C. Z. Laurent, F. Gaill, and O. Gros, Sulphur-oxidizing extracellular bacteria in the gills of Mytilidae associated with wood falls. *FEMS Microbiol. Ecol.*, **63**:3 (2008), 338–49.

Ebbe, B., D. S. M. Billett, A. Brandt, *et al.*, Diversity of abyssal marine life. In A. D. McIntyre (ed.), *Life in the World's Oceans: Diversity, Distribution, and Abundance* (Oxford: Blackwell Publishing Ltd., 2010), pp. 139–60.

Edmonds, H. N., P. J. Michael, E. T. Baker, *et al.*, Discovery of abundant hydrothermal venting on the ultraslow-spreading Gakkel Ridge in the Arctic. *Nature*, **421**:6920 (2003), 252–6.

Ellingsen, K. E., A. Brandt, B. Ebbe, and K. Linse, Diversity and species distribution of polychaetes, isopods and bivalves in the Atlantic sector of the deep Southern Ocean. *Polar Biol.*, **30**:10 (2007), 1265–73.

Erickson, K. L., S. A. Macko, and C. L. Van Dover, Evidence for a chemoautotrophically based food web at inactive hydrothermal vents (Manus Basin). *Deep Sea Res. II*, **56**:19–20 (2009), 1577–85.

Etter, R. J. and J. F. Grassle, Patterns of species-diversity in the deep-sea as a function of sediment particle-size diversity. *Nature*, **360**:6404 (1992), 576–8.

Etter, R. J. and M. A. Rex, Population differentiation decreases with depth in deep-sea gastropods. *Deep Sea Res. I*, **37**:8 (1990), 1251–61.

Etter, R. J., M. A. Rex, M. R. Chase, and J. M. Quattro, Population differentiation decreases with depth in deep-sea bivalves. *Evolution*, **59**:7 (2005), 1479–91.

Falkenhaug, T., A. Gislason, and E. Gaard, *Vertical Distribution and Population Structure of Copepods Along the Northern Mid-Atlantic Ridge* (ICES ASC Helsinki: ICES, 2007).

Forbes, E., *Report on the Mollusca and Radiata of the Aegean Sea*. Report of the British Association for the Advancement of Science for 1842 (1844), pp. 129–33.

de Forges, B. R., J. A. Koslow, and G. C. B. Poore, Diversity and endemism of the benthic seamount fauna in the southwest Pacific. *Nature*, **405**:6789 (2000), 944–7.

Fossa, J. H., P. B. Mortensen, and D. M. Furevik, The deep-water coral *Lophelia pertusa* in Norwegian waters: distribution and fishery impacts. *Hydrobiologia*, **471** (2002), 1–12.

Fossen, I., C. F. Cotton, O. A. Bergstad, and J. E. Dyb, Species composition and distribution patterns of fishes captured by longlines on the Mid-Atlantic Ridge. *Deep Sea Res. II*, **55**:1–2 (2008), 203–17.

Foucher, J. P., G. K. Westbrook, A. Boetius, *et al.*, Structure and drivers of cold seep ecosystems. *Oceanography*, **22**:1 (2009), 92–109.

Fouquet, Y., G. Cherkashov, J. L. Charlou, *et al.*, Serpentine cruise-ultramafic hosted hydrothermal deposits on the Mid-Atlantic Ridge: first submersible studies on Ashadze 1 and 2, Logatchev 2, and Krasnov vent fields. *InterRidge News*, **17** (2008), 15–9.

Francis, R. I. C. C. and M. R. Clark, Sustainability issues for orange roughy fisheries. *Bull. Mar. Sci.*, **76**:2 (2005), 337–51.

Freiwald, A., J. H. Fossa, A. Grehan, J. A. Koslow, and J. M. Roberts, *Cold-water Coral Reefs: Out of Sight – No Longer Out of Mind* (Cambridge, UK: UNEP-WCMC, 2004).

Fujiwara, Y., Whale falls in Japanese waters. *Fossils*, **86** (2009), 1–2.

Fujiwara, Y., M. Kawato, T. Yamamoto, *et al.*, Three-year investigations into sperm whale-fall ecosystems in Japan. *Mar. Ecol.*, **28**:1 (2007), 219–32.

Gaard, E., A. Gislason, T. Falkenhaug, *et al.*, Horizontal and vertical copepod distribution and abundance on the Mid-Atlantic Ridge in June 2004. *Deep Sea Res. II*, **55**:1–2 (2008), 59–71.

Gage, J. D., Why are there so many species in deep-sea sediments? *J. Exp. Mar. Biol. Ecol.*, **200**:1–2 (1996), 257–86.

Gage, J. D. and R. M. May, A dip into the deep seas. *Nature*, **365**:6447 (1993), 609–10.

Gallardo, V. A., Notas sobre la densidad de la fauna bentonica en el sublittoral del norte de Chile. *Gayana (Zool.)*, **10** (1963), 3–15.

Gaston, K. J., P. H. Williams, P. Eggleton, and C. J. Humphries, Large scale patterns of biodiversity: spatial variation in family richness. *Proc. R. Soc. Lond. B Biol. Sci.*, **260**:1358 (1995), 149–54.

Genin, A., Bio-physical coupling in the formation of zooplankton and fish aggregations over abrupt topographies. *J. Mar. Syst.*, **50**:1–2 (2004), 3–20.

Genin, A. and G. W. Boehlert, Dynamics of temperature and chlorophyll structures above a seamount: an oceanic experiment. *J. Mar. Res.*, **43**:4 (1985), 907–24.

Genin, A. and J. Dower, Seamount plankton dynamics. In T. Pitcher, T. Morato, P. J. B. Hart, M. R. Clark, N. Haggan, and R. S. Santos (eds.), *Seamounts: Ecology, Fisheries and Conservation* (Oxford: Wiley-Blackwell, 2007), pp. 85–100.

German, C. R., E. T. Baker, D. P. Connelly, *et al.*, Hydrothermal exploration of the Fonualei Rift and Spreading Center and the Northeast Lau Spreading Center. *Geochem. Geophys. Geosyst.*, 7 (2006), Q11022, doi:10.1029/2006GC001324.

German, C. R., D. R. Yoerger, M. Jakuba, *et al.*, Hydrothermal exploration with the Autonomous Benthic Explorer. *Deep Sea Res. I*, **55**:2 (2008), 203–19.

Gheerardyn, H. and G. Veit-Kohler, Diversity and large-scale biogeography of Paramesochridae (Copepoda, Harpacticoida) in South Atlantic Abyssal Plains and the deep Southern Ocean. *Deep Sea Res. I*, **56**:10 (2009), 1804–15.

Ghosh, A. K. and R. Mukhopadhyay, *Mineral Wealth of the Ocean* (Netherlands: A. A. Balkema, 2000).

Glasby, G. P., Economic geology: lessons learned from deep-sea mining. *Science*, **289**:5479 (2000), 551–3.

Glover, A., G. Paterson, B. Bett, *et al.*, Patterns in polychaete abundance and diversity from the Madeira Abyssal Plain, Northeast Atlantic. *Deep Sea Res. I*, **48**:1 (2001), 217–36.

Glover, A. G. and C. R. Smith, The deep-sea floor ecosystem: current status and prospects of anthropogenic change by the year 2025. *Environ. Conserv.*, **30**:3 (2003), 219–41.

Glover, A. G., C. R. Smith, S. L. Mincks, P. Y. G. Sumida, and A. R. Thurber, Macrofaunal abundance and composition on the West Antarctic Peninsula continental shelf: evidence for a sediment "food bank" and similarities to deep-sea habitats. *Deep Sea Res. II*, **55**:22–23 (2008), 2491–501.

Glover, A. G., C. R. Smith, G. L. J. Paterson, *et al.*, Polychaete species diversity in the central Pacific abyss: local and regional patterns, and relationships with productivity. *Mar. Ecol. Prog. Ser.*, **240** (2002), 157–69.

Gooday, A. J., T. Cedhagen, O. E. Kamenskaya, and N. Cornelius, The biodiversity and biogeography of komokiaceans and other enigmatic foraminiferan-like protists in the deep Southern Ocean. *Deep Sea Res. II*, **54**:16–17 (2007), 1691–719.

Gooday, A. J., L. A. Levin, A. Aranda da Silva, *et al.*, Faunal responses to oxygen gradients on the Pakistan margin: a comparison of foraminiferans, macrofauna and megafauna. *Deep Sea Res. II*, **56**:6–7 (2009), 488–502.

Gooday, A. J., M. G. Malzone, B. J. Bett, and P. A. Lamont, Decadal-scale changes in shallow-infaunal foraminiferal assemblages at the Porcupine Abyssal Plain, NE Atlantic. *Deep Sea Res. II* (in press).

Govenar, B. and C. R. Fisher, Experimental evidence of habitat provision by aggregations of *Riftia pachyptila* at hydrothermal vents on the East Pacific Rise. *Mar. Ecol.*, **28**:1 (2007), 3–14.

Grassle, J. F. and N. J. Maciolek, Deep-sea species richness – regional and local diversity estimates from quantitative bottom samples. *Am. Nat.*, **139**:2 (1992), 313–41.

Gray, J. S., Is deep-sea species diversity really so high? Species diversity of the Norwegian continental shelf. *Mar. Ecol. Prog. Ser.*, **112**:1–2 (1994), 205–9.

Gray, J. S., G. C. B. Poore, K. I. Ugland, *et al.*, Coastal and deep-sea benthic diversities compared. *Mar. Ecol. Prog. Ser.*, **159** (1997), 97–103.

Guinotte, J. M., J. Orr, S. Cairns, *et al.*, Will human-induced changes in seawater chemistry alter the distribution of deep-sea scleractinian corals? *Front. Ecol. Environ.*, **4**:3 (2006), 141–6.

Haase, K. M., A. Koschinsky, S. Petersen, *et al.*, Diking, young volcanism and diffuse hydrothermal activity on the southern Mid-Atlantic Ridge: the Lilliput field at 9°33'S. *Mar. Geol.*, **266**:1–4 (2009), 52–64.

Haedrich, R. L. and G. T. Rowe, Megafaunal biomass in the deep sea. *Nature*, **269**:5624 (1977), 141–2.

Hein, J., Colbalt-rich ferromanganese crusts: global distribution, composition, origin and research activities. Workshop on polymetallic sulphides and cobalt-rich ferromanganese crusts: status and prospects (Jamaica: International Seabed Authority, 2002).

Hessler, R. R. and H. L. Sanders, Faunal diversity in the deep-sea. *Deep Sea Res.*, **14**:1 (1967), 65–78.

Hosia, A., L. Stemmann, and M. Youngbluth, Distribution of net-collected planktonic cnidarians along the northern Mid-Atlantic Ridge and their associations with the main water masses. *Deep Sea Res. II*, **55**:1–2 (2008), 106–18.

Hubbs, C., Initial discoveries of fish faunas on seamounts and offshore banks in the eastern Pacific. *Pac. Sci.*, **12** (1959), 311–16.

Hughes, J. A., T. Smith, F. Chaillan, *et al.*, Two abyssal sites in the Southern Ocean influenced by different organic matter inputs: environmental characterization and preliminary observations on the benthic foraminifera. *Deep Sea Res. II*, **54**:18–20 (2007), 2275–90.

Husebo, A., L. Nottestad, J. H. Fossa, D. M. Furevik, and S. B. Jorgensen, Distribution and abundance of fish in deep-sea coral habitats. *Hydrobiologia*, **471** (2002), 91–9.

Huston, M. A. and D. L. DeAngelis, Competition and coexistence: the effects of resource transport and supply rates. *Am. Nat.*, **144**:6 (1994), 954–77.

IPCC, *Climate Change 2007: Synthesis Report Contribution of Working Groups I, II and III to the Fourth Assessment Report of the Intergovernmental Panel on Climate Change* (Geneva, Switzerland: IPCC. 2007).

Janussen, D., K. R. Tabachnick, and O. S. Tendal, Deep-sea Hexactinellida (Porifera) of the Weddell Sea. *Deep Sea Res. II*, **51**:14–16 (2004), 1857–82.

Johnson, S. B., A. Waren, and R. C. Vrijenhoek, DNA barcoding of *Lepetodrilus* limpets reveals cryptic species. *J. Shellfish. Res.*, **27**:1 (2008), 43–51.

Jorgensen, B. B. and A. Boetius, Feast and famine – microbial life in the deep-sea bed. *Nat. Rev. Microbiol.*, **5**:10 (2007), 770–81.

Kalogeropoulou, V., B. J. Bett, A. J. Gooday, *et al.*, Temporal changes (1989–1999) in deep-sea metazoan meiofaunal assemblages on the Porcupine Abyssal Plain, NE Atlantic. *Deep Sea Res. II*, **57**:15 (2010), 1383–95.

Koschinsky, A., Discovery of new hydrothermal vents on the southern Mid-Atlantic Ridge (4°–10°S) during cruise M68/1. *InterRidge News*, **15** (2006), 9–15.

Koschinsky, A., D. Garbe-Schanberg, S. Sander, *et al.*, Hydrothermal venting at pressure-temperature conditions above the critical point of seawater, 5°S on the Mid-Atlantic Ridge. *Geology*, **36**:8 (2008), 615–8.

Koslow, J. A., G. W. Boehlert, J. D. M. Gordon, *et al.*, Continental slope and deep-sea fisheries: implications for a fragile ecosystem. *ICES J. Mar. Sci.*, **57**:3 (2000), 548–57.

Koslow, J. A., K. Gowlett-Holmes, J. K. Lowry, *et al.*, Seamount benthic macrofauna off southern Tasmania: community structure and impacts of trawling. *Mar. Ecol. Prog. Ser.*, **213** (2001), 111–25.

Krieger, K. J. and B. L. Wing, Megafauna associations with deepwater corals (*Primnoa spp.*) in the Gulf of Alaska. *Hydrobiologia*, **471** (2002), 83–90.

Lambshead, P. J. D., C. J. Brown, T. J. Ferrero, *et al.*, Latitudinal diversity patterns of deep-sea marine nematodes and organic fluxes: a test from the central equatorial Pacific. *Mar. Ecol. Prog. Ser.*, **236** (2002), 129–35.

Lambshead, P. J. D., J. Tietjen, T. Ferrero, and P. Jensen, Latitudinal diversity gradients in the deep sea with special reference to North Atlantic nematodes. *Mar. Ecol. Prog. Ser.*, **194** (2000), 159–67.

Lambshead, P. J. D., J. Tietjen, C. B. Moncrieff, and T. J. Ferrero, North Atlantic latitudinal diversity patterns in deep-sea marine nematode data: a reply to Rex *et al. Mar. Ecol. Prog. Ser.*, **210** (2001), 299–301.

Le Bris, N., M. Zbinden, and F. Gaill, Processes controlling the physico-chemical micro-environments associated with Pompeii worms. *Deep Sea Res. I*, **52**:6 (2005), 1071–83.

Lecroq, B., A. J. Gooday, and J. Pawlowski, Global genetic homogeneity in the deep-sea foraminiferan *Epistominella exigua* (Rotaliida: Pseudoparrellidae). *Zootaxa*, **2096** (2009), 23–32.

Leibold, M. A., M. Holyoak, N. Mouquet, *et al.*, The metacommunity concept: a framework for multi-scale community ecology. *Ecol. Lett.*, **7**:7 (2004), 601–13.

Lenihan, H. S., S. W. Mills, L. S. Mullineaux, *et al.*, Biotic interactions at hydrothermal vents: recruitment inhibition by the mussel *Bathymodiolus thermophilus*. *Deep Sea Res. I*, **55**:12 (2008), 1707–17.

Levin, L. A., Oxygen minimum zone benthos: adaptation and community response to hypoxia. *Oceanogr. Mar. Biol. Annu. Rev.*, **41** (2003), 1–45.

Levin, L. A., Ecology of cold seep sediments: interactions of fauna with flow, chemistry and microbes. *Oceanogr. Mar. Biol. Annu. Rev.*, **43** (2005), 1–46.

Levin, L. A., R. J. Etter, M. A. Rex, *et al.*, Environmental influences on regional deep-sea species diversity. *Annu. Rev. Ecol. Syst.*, **32** (2001), 51–93.

Levin, L. A. and J. D. Gage, Relationships between oxygen, organic matter and the diversity of bathyal macrofauna. *Deep Sea Res. II*, **45**:1–3 (1998), 129–63.

Levin, L. A. and G. F. Mendoza, Community structure and nutrition of deep methane-seep macrobenthos from the North Pacific (Aleutian) Margin and the Gulf of Mexico (Florida Escarpment). *Mar. Ecol.*, **28**:1 (2007), 131–51.

Levin, L. A., G. F. Mendoza, T. Konotchick, and R. Lee, Macrobenthos community structure and trophic relationships within active and inactive Pacific hydrothermal sediments. *Deep Sea Res. II*, **56**:19–20 (2009), 1632–48.

Levin, L. A., C. R. Whitcraft, G. F. Mendoza, J. P. Gonzalez, and G. Cowie, Oxygen and organic matter thresholds for benthic faunal activity on the Pakistan margin oxygen minimum zone (700–1100 m). *Deep Sea Res. II*, **56**:6–7 (2009), 449–71.

Levin, L. A., W. Ziebis, G. F. Mendoza, V. Growney-Cannon, and S. Walther, Recruitment response of methane-seep macrofauna to sulfide-rich sediments: an in situ experiment. *J. Exp. Mar. Biol. Ecol.*, **330**:1 (2006), 132–50.

Linse, K., A. Brandt, J. M. Bohn, *et al.*, Macro- and megabenthic assemblages in the bathyal and abyssal Weddell Sea (Southern Ocean). *Deep Sea Res. II*, **54** (2007), 1848–63.

Lopez-Gonzalez, P. J., J. Bresciani, and M. Conradi, Two new species of *Herpyllobius* Steenstrup & Lutken, 1861 and a new record of *Herpyllobius antarcticus* Vanhoffen, 1913 (parasitic Copepoda) from the Weddell Sea, Antarctica. *Polar Biol.*, **23**:4 (2000), 265–71.

Lopez-Gonzalez, P. J. and J. M. Gili, A new octocoral genus (Cnidaria: Anthozoa) from Antarctic waters. *Polar Biol.*, **23**:7 (2000), 452–8.

Macpherson, E. and C. M. Duarte, Patterns in species richness, size, and latitudinal range of East Atlantic fishes. *Ecography*, **17**:3 (1994), 242–8.

Malyutina, M. V., Revision of *Storthyngura* Vanhoeffen, 1914 (Crustacea: Isopoda: Munnopsididae) with descriptions of three new genera and four new species from the deep South Atlantic. *Org. Divers. Evol.*, **3**:4 (2003), 245–52.

Martins, I., V. Costa, F. M. Porteiro, A. Colaco, and R. S. Santos, Mercury concentrations in fish species caught at Mid-Atlantic Ridge hydrothermal vent fields. *Mar. Ecol. Prog. Ser.*, **320** (2006), 253–8.

May, R. M., Biodiversity – bottoms up for the oceans. *Nature*, **357**:6376 (1992), 278–9.

McClain, C. R., L. Lundsten, M. Ream, J. Barry, and A. DeVogelaere, Endemicity, biogeography, composition, and community structure on a Northeast Pacific seamount. *PLoS ONE*, **4**:1 (2009).

McClain, C. R., M. A. Rex, and R. J. Etter, Patterns in deep-sea macroecology. In J. Witman and K. Roy (eds.), *Marine Macroecology* (Chicago: University of Chicago Press, 2009), pp. 65–100.

McClatchie, S., R. B. Millar, F. Webster, *et al.*, Demersal fish community diversity off New Zealand: Is it related to depth, latitude and regional surface phytoplankton? *Deep Sea Res. I*, **44**:4 (1997), 647–67.

Melchert, B., C. W. Devey, C. R. German, *et al.*, First evidence for high-temperature off-axis venting of deep crustal/mantle heat: the Nibelungen hydrothermal field, southern Mid-Atlantic Ridge. *Earth Planet Sci. Lett.*, **275**:1–2 (2008), 61–9.

Menot, L., J. Galéron, K. Olu, *et al.*, Spatial heterogeneity of macrofaunal communities in and near a giant pockmark area in the deep Gulf of Guinea. *Mar. Ecol.*, **31**:1 (2009), 78–93.

Menot, L., M. Sibuet, R. S. Carney, *et al.*, New perceptions of continental margin biodiversity. In A. D. McIntyre (ed.), *Life in the World's Oceans: Diversity, Distribution, and Abundance* (Oxford: Blackwell Publishing Ltd., 2010), pp. 79–101.

Metaxas, A., Spatial and temporal patterns in larval supply at hydrothermal vents in the northeast Pacific Ocean. *Limnol. Oceanogr.*, **49**:6 (2004), 1949–56.

Miljutin, D. M. and M. A. Miljutina, Deep-sea nematodes of the family Microlaimidae from the Clarion-Clipperton Fracture Zone (North-Eastern Tropic Pacific), with the descriptions of three new species. *Zootaxa*, **2096** (2009), 137–72.

Mittelbach, G. G., C. F. Steiner, S. M. Scheiner, *et al.*, What is the observed relationship between species richness and productivity? *Ecology*, **82**:9 (2001), 2381–96.

Morato, T. and M. R. Clark, Seamount fishes: ecology and life histories. In T. Pitcher, T. Morato, P. J. B. Hart, M. R. Clark, N. Haggan, and R. S. Santos (eds.), *Seamounts: Ecology, Fisheries and Conservation* (Oxford: Wiley-Blackwell, 2007), pp. 170–88.

Mortensen, P. B., L. Buhl-Mortensen, A. V. Gebruk, and E. M. Krylova, Occurrence of deep-water corals on the Mid-Atlantic Ridge based on MAR-ECO data. *Deep Sea Res. II*, **55**:1–2 (2008), 142–52.

Mullineaux, L. S. and S. W. Mills, A test of the larval retention hypothesis in seamount-generated flows. *Deep Sea Res. I*, **44**:5 (1997), 745–70.

Mullineaux, L. S., C. H. Peterson, F. Micheli, and S. W. Mills, Successional mechanism varies along a gradient in hydrothermal fluid flux at deep-sea vents. *Ecol. Monogr.*, **73**:4 (2003), 523–42.

Myers, N., R. A. Mittermeler, C. G. Mittermeler, G. A. B. Da Fonseca, and J. Kent, Biodiversity hotspots for conservation priorities. *Nature*, **403**:6772 (2000), 853–8.

Narayanaswamy, B. E., B. J. Bett, and J. D. Gage, Ecology of bathyal polychaete fauna at an Arctic-Atlantic boundary (Faroe-Shetland Channel, North-east Atlantic). *Mar. Biol. Res.*, **1**:1 (2005), 20–32.

Nozawa, F., H. Kitazato, M. Tsuchiya, and A. J. Gooday, "Live" benthic foraminifera at an abyssal site in the equatorial Pacific nodule province: abundance, diversity and taxonomic composition. *Deep Sea Res. I*, **53**:8 (2006), 1406–22.

O'Hara, T. D., Seamounts: centres of endemism or species richness for Ophiuroids? *Glob. Ecol. Biogeogr.*, **16**:6 (2007), 720–32.

Ondreas, H., M. Cannat, G. Cherkashov, *et al.*, High resolution mapping of the Ashadze and Logachev hydrothermal fields, Mid-Atlantic Ridge 13–15N. American Geophysical Union, Fall Meeting, 2007.

Ondreas, H., K. Olu, Y. Fouquet, *et al.*, ROV study of a giant pockmark on the Gabon continental margin. *Geo.-Mar. Lett.*, **25**:5 (2005), 281–92.

Pailleret, M., T. Haga, P. Petit, *et al.*, Sunken wood from the Vanuatu Islands: identification of wood substrates and preliminary description of associated fauna. *Mar. Ecol.*, **28**:1 (2007), 233–41.

Palanques, A., P. Masque, P. Puig, *et al.*, Anthropogenic trace metals in the sedimentary record of the Llobregat continental shelf and adjacent Foix Submarine Canyon (northwestern Mediterranean). *Mar. Geol.*, **248**:3–4 (2008), 213–27.

Parker, T. and V. Tunnicliffe, Dispersal strategies of the biota on an oceanic seamount: implications for ecology and biogeography. *Biol. Bull.*, **187**:3 (1994), 336–45.

Paterson, G. L. J. and P. J. D. Lambshead, Bathymetric patterns of polychaete diversity in the Rockall Trough, Northeast Atlantic. *Deep Sea Res. I*, **42**:7 (1995), 1199–214.

Paull, C. K., B. Hecker, R. Commeau, *et al.*, Biological communities at the Florida Escarpment resemble hydrothermal vent taxa. *Science*, **226**:4677 (1984), 965–7.

Pawlowski, J., J. Fahrni, B. Lecroq, *et al.*, Bipolar gene flow in deep-sea benthic foraminifera. *Mol. Ecol.*, **16**:19 (2007), 4089–96.

Pierrot-Bults, A. C., A short note on the biogeographic patterns of the Chaetognatha fauna in the North Atlantic. *Deep Sea Res. II*, **55**:1–2 (2008), 137–41.

Poore, G. C. B. and G. D. F. Wilson, Marine species richness. *Nature*, **361**:6413 (1993), 597–8.

Porteiro, F. M. and T. Sutton, Midwater fish assemblages and seamounts. In T. Pitcher, T. Morato, P. J. B. Hart, M. R. Clark, N. Haggan, and R. S. Santos (eds.), *Seamounts: Ecology, Fisheries and Conservation* (Oxford: Wiley-Blackwell, 2007), pp. 101–16.

Priede, I. G., R. Froese, D. M. Bailey, *et al.*, The absence of sharks from abyssal regions of the world's oceans. *Proc. R. Soc. Lond. B Biol. Sci.*, **273**:1592 (2006), 1435–41.

Priede, I. G., J. A. Godbold, N. J. King, *et al.*, Deep-sea demersal fish species richness in the Porcupine Seabight, NE Atlantic Ocean: global and regional patterns. *Mar. Ecol.*, **31**:1 (2010), 247–60.

Probert, P. K., S. Christiansen, K. M. Gjerde, S. Gubbay, and R. S. Santos, Mangement and conservation of seamounts. In T. Pitcher, T. Morato, P. J. B. Hart, M. Clark, N. Haggan, and R. S. Santos (eds.), *Seamounts: Ecology, Fisheries and Conservation* (Oxford: Wiley-Blackwell, 2007), pp. 442–75.

Ramirez-Llodra, E., J. B. Company, and F. Sarda, Megabenthic diversity patterns and community structure of the Blanes Submarine Canyon and adjacent slope in the Northwestern Mediterranean: a human overprint? *Mar. Ecol.*, **32** (2010), 167–82.

Raupach, M. J., C. Held, and J. W. Wagele, Multiple colonization of the deep sea by the Asellota (Crustacea: Peracarida: Isopoda). *Deep Sea Res. II*, **51**:14–16 (2004), 1787–95.

Raupach, M. J., C. Mayer, M. Malyutina, and J. W. Wagele, Multiple origins of deep-sea Asellota (Crustacea: Isopoda) from shallow waters revealed by molecular data. *Proc. R. Soc. Lond. B Biol. Sci.*, **276**:1658 (2009), 799–808.

Rex, M. A., Community structure in the deep-sea benthos. *Annu. Rev. Ecol. Syst.*, **12** (1981), 331–53.

Rex, M. A., Geographic patterns of species diversity in deep-sea benthos. In G. T. Rowe (ed.), *The Sea, Vol. 8. Deep Sea Biology* (New York: Wiley, 1983), pp. 453–72.

Rex, M. A., J. A. Crame, C. T. Stuart, and A. Clarke, Large-scale biogeographic patterns in marine mollusks: a confluence of history and productivity? *Ecology*, **86**:9 (2005), 2288–97.

Rex, M. A., R. J. Etter, J. S. Morris, *et al.*, Global bathymetric patterns of standing stock and body size in the deep-sea benthos. *Mar. Ecol. Prog. Ser.*, **317** (2006), 1–8.

Rex, M. A., C. R. McClain, N. A. Johnson, *et al.*, A source-sink hypothesis for abyssal biodiversity. *Am. Nat.*, **165**:2 (2005), 163–78.

Rex, M. A., C. T. Stuart, and G. Coyne, Latitudinal gradients of species richness in the deep-sea benthos of the North Atlantic. *Proc. Natl. Acad. Sci. USA*, **97**:8 (2000), 4082–5.

Rex, M. A., C. T. Stuart, and R. J. Etter, Do deep-sea nematodes show a positive latitudinal gradient of species diversity? The potential role of depth. *Mar. Ecol. Prog. Ser.*, **210** (2001), 297–8.

Rex, M. A., C. T. Stuart, R. R. Hessler, *et al.*, Global-scale latitudinal patterns of species-diversity in the deep-sea benthos. *Nature*, **365**:6447 (1993), 636–9.

Richter, T. O., H. C. de Stigter, W. Boer, C. C. Jesus, and T. C. E. van Weering, Dispersal of natural and anthropogenic lead through submarine canyons at the Portuguese margin. *Deep Sea Res. I*, **56**:2 (2009), 267–82.

Ricketts, E. R., J. P. Kennett, T. M. Hill, and J. P. Barry, Effects of carbon dioxide sequestration on California margin deep-sea foraminiferal assemblages. *Mar. Micropaleontol.*, **72**:3–4 (2009), 165–75.

Roberts, J. M., A. J. Wheeler, and A. Freiwald, Reefs of the deep: the biology and geology of cold-water coral ecosystems. *Science*, **312**:5773 (2006), 543–7.

Rodriguez, E. and P. J. Lopez-Gonzalez, *Stephanthus antarcticus*, a new genus and species of sea anemone (Actiniaria, Haloclavidae) from the South Shetland Islands, Antarctica. *Helgol. Mar. Res.*, **57**:1 (2003), 54–62.

Rosenzweig, M. L., Species-diversity gradients – we know more and less than we thought. *J. Mammal.*, **73**:4 (1992), 715–30.

Ross, S. W. and A. M. Quattrini, The fish fauna associated with deep coral banks off the southeastern United States. *Deep Sea Res. I*, **54**:6 (2007), 975–1007.

Rowden, A. A., M. R. Clark, and I. C. Wright, Physical characterisation and a biologically focused classification of "seamounts" in the New Zealand region. *NZ J. Mar. Freshwat. Res.*, **39**:5 (2005), 1039–59.

Rowe, G. T., Benthic biomass and surface productivity. In J. D. Costlow (ed.), *Fertility of the Sea* (New York: Gordon and Breach, 1971), pp. 441–54.

Rowe, G. T. and R. J. Menzies, Zonation of large benthic invertebrates in the deep-sea off the Carolinas. *Deep Sea Res.*, **16**:5 (1969), 531–7.

Roy, K. O. L., J. C. Caprais, A. Fifis, *et al.*, Cold-seep assemblages on a giant pockmark off West Africa: spatial patterns and environmental control. *Mar. Ecol.*, **28**:1 (2007), 115–30.

Roy, K., D. Jablonski, J. W. Valentine, and G. Rosenberg, Marine latitudinal diversity gradients: tests of causal hypotheses. *Proc. Natl. Acad. Sci. USA*, **95**:7 (1998), 3699–702.

Ruhl, H. A. and K. L. Smith, Shifts in deep-sea community structure linked to climate and food supply. *Science*, **305** (2004), 513–15.

Samadi, S., L. Bottan, E. Macpherson, B. R. De Forges, and M. C. Boisselier, Seamount endemism questioned by the geographic distribution and population genetic structure of marine invertebrates. *Mar. Biol.*, **149**:6 (2006), 1463–75.

Sancho, G., C. R. Fisher, S. Mills, *et al.*, Selective predation by the zoarcid fish *Thermarces cerberus* at hydrothermal vents. *Deep Sea Res. I*, **52**:5 (2005), 837–44.

Sanders, H. L., Marine benthic diversity – a comparative study. *Am. Nat.*, **102**:925 (1968), 243.

Sanders, H. L. and R. R. Hessler, Ecology of deep-sea benthos. *Science*, **163**:3874 (1969), 1419.

Schlacher, T., A. Williams, F. Althaus, and M. A. Schlacher-Hoenlinger, High-resolution seabed imagery as a tool for biodiversity conservation planning on continental margins. *Mar. Ecol.*, **31**:1 (2010), 200–21.

Schuller, M. and B. Ebbe, Global distributional patterns of selected deep-sea Polychaeta (Annelida) from the Southern Ocean. *Deep Sea Res. II*, **54** (2007), 1737–51.

Sedlacek, L., D. Thistle, K. R. Carman, J. W. Fleeger, and J. P. Barry, Effects of carbon dioxide on deep-sea harpacticoids revisited. *Deep Sea Res. I*, **56**:6 (2009), 1018–25.

Sellanes, J., C. Neira, E. Quiroga, and N. Teixido, Diversity patterns along and across the Chilean margin: a continental slope encompassing oxygen gradients and methane seep benthic habitats. *Mar. Ecol.*, **31**:1 (2010), 111–24.

Sellanes, J., E. Quiroga, and C. Neira, Megafauna community structure and trophic relationships at the recently discovered Concepcíon Methane Seep Area, Chile, ~36°S. *ICES J. Mar. Sci.*, **65**:7 (2008), 1102–11.

Sibuet, M. and K. Olu, Biogeography, biodiversity and fluid dependence of deep-sea cold-seep communities at active and passive margins. *Deep Sea Res. II*, **45**:1–3 (1998), 517–67.

Sibuet, M. and A. Vangriesheim, Deep-sea environment and biodiversity of the West African Equatorial margin. *Deep Sea Res. II*, **56**:23 (2009), 2156–68.

Skadsheim, A., J. F. Borseth, A. Bjornstad, *et al.*, Hydrocarbons and chemicals: potential effects and monitoring in the deep sea. In S. L. Armsworthy, P. J. Cranford, and K. Lee (eds.), *Offshore Oil and Gas Environmental Effects Monitoring: Approaches and Technologies* (Columbus, OH: Battelle Press, 2005), p. 631.

Skov, H., T. Gunnlaugsson, W. P. Budgell, *et al.*, Small-scale spatial variability of sperm and sei whales in relation to oceanographic and topographic features along the Mid-Atlantic Ridge. *Deep Sea Res. II*, **55**:1–2 (2008), 254–68.

Smith, C. R. and A. R. Baco, Ecology of whale falls at the deep-sea floor. *Oceanogr. Mar. Biol. Annu. Rev.*, **41** (2003), 311–54.

Smith, C. R., F. C. De Leo, A. F. Bernardino, A. K. Sweetman, and P. M. Arbizu, Abyssal food limitation, ecosystem structure and climate change. *Trends Ecol. Evol.*, **23**:9 (2008), 518–28.

Smith, C. R., D. J. Hoover, S. E. Doan, *et al.*, Phytodetritus at the abyssal seafloor across 10° of latitude in the central equatorial Pacific. *Deep Sea Res. II*, **43**:4–6 (1996), 1309–38.

Smith, C. R., H. Kukert, R. A. Wheatcroft, P. A. Jumars, and J. W. Deming, Vent fauna on whale remains. *Nature*, **341**:6237 (1989), 27–8.

Smith, C. R., L. A. Levin, J. A. Koslow, P. Tyler, and A. G. Glover, The near future of the deep seafloor ecosystems. In N. Polunin (ed.), *Aquatic Ecosystems: Trends and Global Prospects* (Cambridge: Cambridge University Press, 2008), pp. 334–51.

Smith, C. R., G. L. J. Paterson, P. J. D. Lambshead, *et al.*, *Biodiversity, Species Ranges, and Gene Flow in the Abyssal Pacific Nodule Province: Predicting and Managing the Impacts of Deep Seabed Mining* (Kingston, Jamaica: International Seabed Authority, 2008).

Smith Jr., K. L., Benthic community respiration in the N.W. Atlantic Ocean: in situ measurements from 40 to 5200 m. *Mar. Biol.*, **47**:4 (1978), 337–47.

Smith, K. L., H. A. Ruhl, B. J. Bett, *et al.*, Climate, carbon cycling, and deep-ocean ecosystems. *Proc. Natl. Acad. Sci. USA*, **106**:46 (2009), 19211–8.

Smith, P. J., S. M. McVeagh, J. T. Mingoia, and S. C. France, Mitochondrial DNA sequence variation in deep-sea bamboo coral (Keratoisidinae) species in the southwest and northwest Pacific Ocean. *Mar. Biol.*, **144**:2 (2004), 253–61.

Snelgrove, P. V. R. and C. R. Smith, A riot of species in an environmental calm: the paradox of the species-rich deep-sea floor. *Oceanogr. Mar. Biol. Annu. Rev.*, **40** (2002), 311–42.

Solé, M., C. Porte, and J. Albaigés, Hydrocarbons, PCBs and DDT in the NW Mediterranean deep-sea fish *Mora moro*. *Deep Sea Res. I*, **48**:2 (2001), 495–513.

Soh, W., Transport processes deduced from geochemistry and the void ratio of surface core samples, deep sea Sagami Bay, central Japan. *Prog. Oceanogr.*, **57**:1 (2003), 109–24.

Stemmann, L., A. Hosia, M. J. Youngbluth, *et al.*, Vertical distribution (0–1000 m) of macrozooplankton, estimated using the Underwater Video Profiler, in different hydrographic regimes along the northern portion of the Mid-Atlantic Ridge. *Deep Sea Res. II*, **55**:1–2 (2008), 94–105.

Stockley, B., G. Menezes, M. R. Pinho, and A. D. Rogers, Genetic population structure in the black-spot sea bream (*Pagellus bogaraveo* Brannich, 1768) from the NE Atlantic. *Mar. Biol.*, **146**:4 (2005), 793–804.

Stocks, K. I., G. W. Boehlert, and J. F. Dower, Towards an international field programme on seamounts within the Census of Marine Life. *Arch. Fish. Mar. Res.*, **51**:1–3 (2004), 320–7.

Stocks, K. I., C. Condit, X. Qian, P. E. Brewin, and A. Gupta, Bringing together an ocean of information: an extensible data integration framework for biological oceanography. *Deep Sea Res. II*, **56**:19–20 (2009), 1804–11.

Stocks, K. I. and P. J. B. Hart, Biogeography and biodiversity of seamounts. In T. Pitcher, T. Morato, P. J. B. Hart, M. R. Clark, N. Haggan, and R. S. Santos (eds.), *Seamounts: Ecology, Fisheries and Conservation* (Oxford: Wiley-Blackwell, 2007), pp. 255–81.

Storelli, M. M., S. Losada, G. O. Marcotrigiano, *et al.*, Polychlorinated biphenyl and organochlorine pesticide contamination signatures in deep-sea fish from the Mediterranean Sea. *Environ. Res.*, **109**:7 (2009), 851–6.

Stramma, L., G. C. Johnson, J. Sprintall, and V. Mohrholz, Expanding oxygen-minimum zones in the tropical oceans. *Science*, **320**:5876 (2008), 655–8.

Sutton, T. T., F. M. Porteiro, M. Heino, *et al.*, Vertical structure, biomass and topographic association of deep-pelagic fishes in relation to a mid-ocean ridge system. *Deep Sea Res. II*, **55**:1–2 (2008), 161–84.

Thiel, H., Anthropogenic impacts in the deep sea. In P. Tyler (ed.), *Ecosystems of the Deep Ocean* (Amsterdam: Elsevier, 2003), pp. 427–72.

Tselepides, A. and N. Lampadariou, Deep-sea meiofaunal community structure in the Eastern Mediterranean: are trenches benthic hotspots? *Deep Sea Res. I*, **51**:6 (2004), 833–47.

Tunnicliffe, V., A. G. McArthur, and D. McHugh, A biogeographical perspective of the deep-sea hydrothermal vent fauna. *Adv. Mar. Biol.*, **34** (1998). 353–442.

Turley, C. M., J. M. Roberts, and J. M. Guinotte, Corals in deep-water: will the unseen hand of ocean acidification destroy cold-water ecosystems? *Coral Reefs*, **26**:3 (2007), 445–8.

Turnipseed, M., K. E. Knick, R. N. Lipcius, J. Dreyer, and C. L. Van Dover, Diversity in mussel beds at deep-sea hydrothermal vents and cold seeps. *Ecol. Lett.*, **6**:6 (2003), 518–23.

Tyler, P., C. German, and V. Tunnicliffe, Biologists do not pose a threat to deep-sea vents. *Nature*, **434**:7029 (2005), 18–.

Tyssebotn, I. M., Contamination in deep sea fish: toxic elements, dioxins, furans and dioxin-like PCBs in the orange roughy. MSc thesis, University of Bergen (2008).

Van Dover, C., *The Ecology of Deep-sea Hydrothermal Vents* (Princeton: Princeton University Press, 2000).

Van Dover, C. L., Biodiversity at deep-sea vent and intertidal mussel beds. *Am. Zool.*, **39**:5 (1999), 676.

Van Dover, C. L., C. R. German, K. G. Speer, L. M. Parson, and R. C. Vrijenhoek, Marine biology – evolution and biogeography of deep-sea vent and seep invertebrates. *Science*, **295**:5558 (2002), 1253–7.

Van Gaest, A. L., C. M. Young, J. J. Young, A. R. Helms, and S. M. Arellano, Physiological and behavioral responses of *Bathynerita naticoidea* (Gastropoda: Neritidae) and *Methanoaricia dendrobranchiata* (Polychaeta: Orbiniidae) to hypersaline conditions at a brine pool cold seep. *Mar. Ecol.*, **28**:1 (2007), 199–207.

Van Gaever, S., L. Moodley, F. Pasotti, *et al.*, Trophic specialisation of metazoan meiofauna at the Hakon Mosby mud volcano: fatty acid biomarker isotope evidence. *Mar. Biol.*, **156**:6 (2009), 1289–96.

Van Gaever, S., K. Olu, S. Derycke, and A. Vanreusel, Metazoan meiofaunal communities at cold seeps along the Norwegian margin: influence of habitat heterogeneity and evidence for connection with shallow-water habitats. *Deep Sea Res. I*, **56**:5 (2009), 772–85.

Vanreusel, A., G. Fonesca, R. Danovaro, *et al.*, The importance of deep-sea habitat heterogeneity for global nematode diversity. *Mar. Ecol.*, **31**:1 (2010), 6–20.

Vecchione, M., O. A. Bergstad, I. Byrkjedal, *et al.*, Biodiversity patterns and processes on the Mid-Atlantic Ridge. In A. D. McIntyre (ed.), *Life in the World's Oceans: Diversity, Distribution and Abundance* (Oxford: Blackwell Publishing Ltd., 2010), pp. 103–21.

Vecchione, M. and R. E. Young, The squid family Magnapinnidae (Mollusca: Cephalopoda) in the Atlantic Ocean, with a description of a new species. *Proc. Biol. Soc. Wash.*, **119**:3 (2006), 365–72.

Veillette, J., J. Sarrazin, A. J. Gooday, *et al.*, Ferromanganese nodule fauna in the Tropical North Pacific Ocean: species richness, faunal cover and spatial distribution. *Deep Sea Res. I*, **54**:11 (2007), 1912–35.

Vermeeren, H., A. Vanreusel, and S. Vanhove, Species distribution within the free-living marine nematode genus *Dichromadora* in the Weddell Sea and adjacent areas. *Deep Sea Res. II*, **51**:14–16 (2004), 1643–64.

Vetter, E. W., C. R. Smith, and F. C. DeLeo, Hawaiian hotspots: enhanced megafaunal abundance and diversity in submarine canyons on the oceanic islands of Hawaii. *Mar. Ecol.*, **31**:1 (2010), 183–99.

Webb, T. J., E. Vanden Berghe, and R. O'Dor, Biodiversity's big wet secret: the global distribution of marine biological records reveals chronic under-exploration of the deep pelagic ocean. *PLoS ONE* (2010), e10223, doi: 10.1371/journal.pone.0010223.

Wei, C. L. and G. T. Rowe, Faunal zonation of large epibenthic invertebrates off North Carolina revisited. *Deep Sea Res. II*, **56**:19–20 (2009), 1830–3.

Wheeler, A. J., B. J. Bett, D. S. M. Billett, D. G. Masson, and D. Mayor, The impact of demersal trawling on NE Atlantic deep-water coral habitats: the case of the Darwin Mounds, UK. *Benthic Habit. Eff. Fish.*, **41** (2005), 807–17.

White, M., I. Bashmachnikov, J. Aristegui, and A. Martins, Physical processes and seamount productivity. In T. Pitcher, T. Morato, P. J. B. Hart, M. R. Clark, N. Haggan, and R. S. Santos (eds.), *Seamounts: Ecology, Fisheries and Conservation* (Oxford: Wiley-Blackwell, 2007), pp. 65–84.

Williams, A., N. J. Bax, R. J. Kloser, *et al.*, Australia's deep-water reserve network: implications of false homogeneity for classifying abiotic surrogates of biodiversity. *ICES J. Mar. Sci.*, **66**:1 (2009), 214–24.

Witte, U., F. Wenzhofer, S. Sommer, *et al.*, In situ experimental evidence of the fate of a phytodetritus pulse at the abyssal sea floor. *Nature*, **424**:6950 (2003), 763–6.

Wolff, T., Macrofaunal utilization of plant remains in the deep-sea. *Sarsia*, **64**:1–2 (1979), 117.

World Ocean Atlas (database on the Internet) 2005.

Young, R. E., A. Lindgren, and M. Vecchione, *Mastigoteuthis microlucens*, a new species of the squid family Mastigoteuthidae (Mollusca: Cephalopoda). *Proc. Biol. Soc. Wash.*, **121**:2 (2008), 276–82.

Young, R. E., M. Vecchione, and U. Piatkowski, *Promachoteuthis sloani*, a new species of the squid family Promachoteuthidae (Mollusca: Cephalopoda). *Proc. Biol. Soc. Wash.*, **119**:2 (2006), 287–92.

Youngbluth, M., T. Sornes, A. Hosla, and L. Stemmann, Vertical distribution and relative abundance of gelatinous zooplankton, in situ observations near the Mid-Atlantic Ridge. *Deep Sea Res. II*, **55**:1–2 (2008), 119–25.

Zeidberg, L. D. and B. H. Robison, Invasive range expansion by the Humboldt squid, *Dosidicus gigas*, in the eastern North Pacific. *Proc. Natl. Acad. Sci. USA*, **104**:31 (2007), 12948–50.

CHAPTER

9

Changing ocean

Natural change in Planet Ocean

Tides change the ocean daily, and cause many of the patterns around the
ocean rim. While we hardly notice a small change in tide along the Australian
coast, tourists flock to see a 16-meter rise and fall in the Bay of Fundy, floating
and then grounding fishing boats. El Niño affects the weather far and wide,
but it throttles upwelling of cold, nutrient-rich water off Peru that feeds
large fish populations, which in turn sustain abundant seabirds. The
ocean changes for myriad reasons.

Preserved Danish fish bones show species composition changed during
a warm period some 6,000 to 9,000 years ago. Baltic Sea herring, *Clupea
harengus*, and other species declined in cold waters in the late 1600s, as did
salmon, halibut, and cod in the Barents and White Seas. Preserved fish scales
from sediments off the California coast show fluctuating anchovy and sardine
over the past 1,700 years, long before humans arrived. Fisheries' biologists
reconstructed a Danish fjord fishery from 1667 to 1860 and concluded that
overfishing collapsed herring, but a major storm that opened the fjord to
the North Sea and saltwater intrusion lowered eel and whitefish numbers.
Thus, some collapses were natural, and another avoidable.

The ocean changes naturally, but now we humans too drive change.
What would the ocean look like if we had not? We don't know because early
fishermen were the first to sample the ocean, and their sampling expanded
quickly and efficiently as small fishing boats propelled by human muscle

or wind were replaced by powerful trawlers that haul heavy fishing gear across the seafloor and through the water. Scientists seek compelling proof to understand change, but data that show the ocean before exploitation simply don't exist. The analogy of a "human footprint" on the ocean is appropriate because footprints tell you someone was there, but not what they changed.

Changing oceans and changing marine life extend well beyond our attention spans or the decade of the Census. We suffer "shifting baseline syndrome," where our lifetime experience colors our expectation of a "healthy" ocean to one that may differ from historical reality. I was surprised when (the now deceased) Ransom Myers, a leader within the Census and a master of global data integration, noted that walrus were hunted to extinction in Newfoundland many years before I was born. Although a native resident who studies Newfoundland waters, I did not know walrus had ever lived there and had suffered my own shifting baseline syndrome.

We can see recent change, but must infer past changes from archaeological sources and other historical clues rather than surveys of the present ocean. Accordingly, the Census' *History of Marine Animal Populations* (HMAP), established the new discipline of environmental history. Tim Smith, one of the founding leaders of the HMAP project, describes it as an attempt to correct "historical myopia." Using numbers standardized for historical fishing effort, biodiversity metrics, and statistical models, HMAP asked how the diversity, distribution, and abundance of marine life in Planet Ocean changed over thousands of years. They worked closely with the *Future of Marine Animal Populations* (FMAP) project, which focused on recent changes in ocean life, and in some cases how those changes may foretell the future.

Early remains and images of oceans past

Archaeological evidence provides uneven "glimpses of the past." Prehistoric man drew seals and auks on cave walls in the Grotto Cosquer in France 18,000 years ago. Boats from Egypt in 4000 BC and China and Kuwait from 7,000 to 7,500 years ago tell us ocean transport began early, but we are less certain when boats expanded fishing effort. In 2400 BC, in the oldest known civilization, Sumer, hundreds of fishermen caught more fish in the Tigris-Euphrates with hook and line than they could consume. Prior to nets and fishing boats, ancient fishers relied on hooks and worked from shore, suggesting fish were a minor dietary source. But before accepting that view, remember that nets are poorly preserved, and some modern fishermen catch many fish while standing on shore.

When did boats first extend fishing nets, increase catches, and when did salting preserve the catch for storage and transport? Before salting, which dates back to the seventh century BC and possibly to 2400 BC, fishermen had no way to preserve their catch. Fishing very early in the morning before the midday heat bought a little time, but still limited what they could catch and deliver.

Images of tuna and sturgeon on early Greek coins (about 400 BC) suggest they were part of that culture. Imagery on shields from 400 BC excavated from the northern Black Sea show that early nomads fished, a finding seconded by abundant fish bones in household middens. By the early first century BC, Romans operated fish-processing facilities in Spain, Portugal, and Morocco. Many facilities were built near coastal "pinch points" to funnel migrating or spawning tuna, mackerel, and eels into nets. We may never know when these activities had any appreciable impact on fish populations, but they mark the beginnings of "commercial" fishing, and wide distribution of fish products through Western Europe by the first century AD, as Roman influence spread.

The best-documented changes are for fishes. Shells of clams and other mollusks preserve well on land, however, and demonstrate early human exploitation of islands and coasts. Large mollusks (*megamollusks*) have been used as food and ornaments for over 450,000 years, and have contributed to human demography and migration for 150,000 years. Between 1200 and 1500 AD in the New World off Venezuela, humans harvested more than five million conches, leaving a massive pre-Columbian midden of queen conch (Figure 9.1). Still, humans did not depend enough on mollusks to drive them to extinction.

Paintings tell a story of change in Northern European waters; Frans Snyders' paintings of fish markets at the turn of the seventeenth century show an abundance and variety of fish no longer seen 300 years later in European waters. But images, bones, and shells give glimpses of what humans fished rather than their impact, and written information is more telling.

Early writings about oceans past

HMAP also used ships' logs, surveyors' and pirates' diaries, tax records, and medieval cookbooks and restaurant menus to look back in time at how humans have changed the ocean. Early Greek writing confirms some

Figure 9.1 A jetty of mollusks past.
Middens of queen conch (*Strombus gigas*) date from 1200 to 1500 AD on La Pelona Island, Los Roques Archipelago, Venezuela, forming a jetty. The size of the shell jetty shows the importance of mollusks to humans in that era.

archaeological conclusions. The Hippocratic *Regimen II* of about 400 BC refers repeatedly to salted fish. The Greek writer Oppian's *Halieutika* was the first treatise on fishing, and describes fishermen using boats in the first century AD. Pliny the Elder's written fish recipes tell us fish was a staple food more than two millennia ago at the time of the Roman Empire. Perhaps *garum*, the fermented fish sauce that was a Roman staple and transported throughout Western Europe, was the antiquarian version of McDonald's "special sauce."

As western civilization and human populations spread, fishermen and the food they landed grew in numbers. By 1000 AD in the North Sea, migrating salmon and eel, and later cod, ling, and herring caught by boats, were major staples. In the 1200s, the Baltic Sea herring fishery was a major industry. Giovanni Caboto, sent by England in 1497 to find New World fish to replace declining European stocks, described cod in Newfoundland waters so plentiful they could be captured in baskets.

Unnatural changes in Planet Ocean

Planet Ocean continued to feed more and more people, but fishing
expansion and intensification was taking a toll on the natural environment.
Impacts appeared centuries ago, even before industrial fishing in recent decades.
Sparse populations of indigenous people probably harvested, but had little
impact on, shellfish and fishes. But the evolution of commerce moved fishing
sequentially from estuaries to shelf to open ocean. Depletion of estuaries and
densely populated nearshore regions dates back hundreds to thousands of
years, shelf fisheries began precipitous declines within the last 50 years,
and deep-sea and oceanic species declined mostly in the last few decades.

Some took note long ago. In 1376, British fishers petitioned the
British Parliament that trawling was destroying "the living slym and
underwater plants," and that catches of smaller fish were declining. But
fishing continued to expand. By the fourteenth century, sturgeon from
rivers flowing to the Wadden Sea had disappeared from markets, becoming
a food only for the aristocracy. By the sixteenth century, annual Danish fish
catches were at 35,000 tons, and Dutch fishermen landed 75,000 tons annually
before major declines in the 1700s. Poor fishing practices collapsed the Danish
herring fishery by 1830, and it has never fully recovered. As markets in
Europe became more global around that time, seabirds, marine mammals,
and reptiles declined in many areas. When motorized vessels were launched
around 1900, declines in commercial salmon and cod in Denmark quickly
followed. The increased cost of running a fishing boat simultaneously
created a need to catch more fish.

Marine fisheries shaped patterns of trade and human settlement of
North America from the 1500s onward. Early wars over fishing rights, the
trade routes that ensued, and patterns of human settlement closely linked to the
development of seventeenth and eighteenth century cod fisheries in the western
Atlantic. Increased exploitation led to serious decline that set the stage for stock
collapses in recent decades; many have still not recovered. The past view of
resilient fish stocks easily recovered with short-term closures is quickly fading.

Recent change in Planet Ocean

Where HMAP ends, FMAP picks up and carries the story of ocean
change through the present. Australia is unique because baseline data

pre-date exploitation, but the story they tell is familiar. Australian fisheries only began to ramp up at the start of the twentieth century, but motorized boats and gear were already available, and by the 1930s declining catch pushed fishermen into deeper and deeper waters in pursuit of fish. Eventually, specially designed fishing gear and electronic devices to locate fishes on the seafloor would soon exploit areas whose rugged bottoms had previously protected them from fishing. Orange roughy were quickly and efficiently fished to commercial extinction on distant seamounts with much collateral damage to other species. Although elaborate fishing gear accelerates rapid decline, it is not a prerequisite; Jamaicans fished coral reef fishes beyond sustainable levels a century ago with simple fishing gear such as hook and line.

Recognizing declines in fish stocks, fisheries science grew quickly after World War II with widely adopted fisheries models of Ricker, Beverton and Holt, and others. Although these tools helped in some cases, they demand good data that extend back over time. But *time series* data are difficult to interpret because fishing effort changes in space and time, as do discards, and erroneous catch reporting. Even different vessels using similar bottom trawls catch at different rates. In some cases fisheries management has produced recoveries, but many fisheries remain depleted long after we try to manage them.

Good knowledge helps. New analytical approaches can determine when *depensation* occurs, where a population drops so low that recovery is near impossible. In an example of the power of knowledge, new analysis of coho salmon and alewife populations ruled out depensation so fisheries managers could opt for less extreme measures than completely shutting down a fishery.

Nevertheless, depletions continue, one species after another. *Serial depletion* fishes down one species, and then moves on to deplete another, eventually "fishing down the food web." In Newfoundland, cod followed haddock and now crab and shrimp follow cod. Fishermen move through multiple species, often of decreasing economic value, and begin to target "underutilized species" that are little studied and therefore even more difficult to manage.

Factory freezer trawlers scoop up tons of fish in a single trawl, accelerating declines. Industrialized fisheries today can reduce community biomass by 80% in 15 years. We are far more efficient ocean predators than we were 100 years ago, but also far more wasteful as we throw tons of discard over the side. Evolving technologies and faster boats depleted whale species in series, beginning with slow, inshore species like right whales before moving to faster offshore species like sperm whales. Again we fish the most valuable or accessible before moving on to less valuable or hard-to-catch species.

Declining fisheries bring social and economic change, mostly for the worse. The fishing industry directly employs some 200 million people and provides some 5% of the total protein and 20% of animal protein in human diets globally, and much more in poorer coastal nations. Tuna remains valuable; a single Pacific bluefin tuna, *Thunnus orientalis*, fetched $174,000 US in 2001! Because the stakes around the world are so high, reports on overfishing attract attention and controversy. We now reluctantly see global-scale fisheries collapses. "Perverse incentives" to upgrade boats or supplement seasonal employment save jobs, but escalate exploitation, and although some argue that removing these subsidies would eliminate marginal fisheries and allow stocks to rebuild, this is a cruel solution for fishermen with mortgages and hungry families.

What have we lost?

In 1883, naturalist Thomas Henry Huxley optimistically stated that "I believe, then, that the cod fishery, the herring fishery, the pilchard fishery, the mackerel fishery, and probably all the great sea fisheries, are inexhaustible; that is to say, that nothing we do seriously affects the number of the fish. And any attempt to regulate these fisheries seems consequently, from the nature of the case, to be useless." His contemporary, Edwin Lankester had a different view. "If man removes a large proportion of these fish from the areas which they inhabit, the natural balance is upset." The voice of the industrial boom was loud, as it often is today, and many changes would follow. Consider what we lost that damped the optimism of Huxley's view and tipped history toward Lankester's fear of humanity's power to unbalance a system as vast as Planet Ocean.

Great auks and Atlantic gray whales were extinct in the North Sea by the late medieval period and disappeared globally not long after. New England and Newfoundland (Figure 9.2) fishermen once caught 1- to 2-meter cod that were sometimes bigger than them, and Chesapeake oyster beds produced 30-centimeter oysters that had to be cut in pieces to be eaten. In the Wadden Sea, some 20% of the *macrobiota*, those organisms visible to the naked eye,

Figure 9.2 Small boy, big cod.
Today fishermen rarely catch cod like this 25-kilogram prize caught near Battle Harbour, Labrador, just north of the island of Newfoundland, in the early 1900s.

disappeared or seriously declined over the last 1,000 years. Abundant large fishes, sea turtles, and marine mammals that once populated coral reefs are ignored in modern textbooks because they were gone before researchers arrived in the 1950s. Ships' logs show that cod biomass estimates for Canada's Scotian shelf declined some 96% from 1,260,000 metric tons in 1852 to 50,000 metric tons in 2005.

Archaeological data from eastern Canada show two centuries of decline, mostly from overfishing. European arrival in the 1700s caused declines by the 1800s in *diadromous fishes*, which use both fresh and salt water, followed by declines in groundfish and the extinction of six bird and three marine mammal species from the region by 1900. These declines set the stage for collapse of groundfish by the 1970s that would be mirrored in other areas

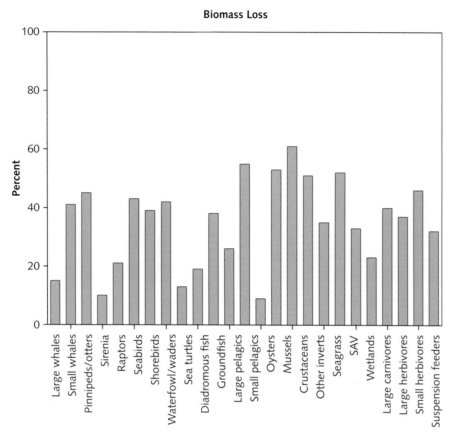

Figure 9.3 Life lost.
The total abundance of major groups of marine life declined by about 10 to 60% in 12 estuaries and coastal seas around the world since human arrival. SAV refers to submerged aquatic vegetation.

of Atlantic Canada in the decades that followed and the grim state of many fisheries today. Since exploitation began, the weight or biomass of 256 species of large marine mammals, birds, reptiles, and fishes declined 89%, and individual species declined by 11–100% (Figure 9.3). The total weight of large predatory fish has fallen to 10% of pre-industrial fishing levels. These changes mirror declines in biodiversity, the array of species present. Our knowledge of highly valued commercial species is far greater than our knowledge of others, so many biodiversity changes go unobserved.

HMAP reconstructed abundance and distribution of whales from logs of whaling ships to produce maps that tell of 70,000 whale encounters and locations during the nineteenth century. These data show frequent and

Figure 9.4 Abundant North Sea bluefin tuna in oceans past.
Northern bluefin tuna, *Thunnus thynnus*, for sale in the fish auction hall at Skagen, Denmark, about 1946. Today, their numbers are much diminished.

alarming *extirpations*, or local eliminations of marine mammals. The oceans between east Australia and New Zealand teemed with 27,000 southern right whales, roughly 30 times as many as today. Using genetic diversity and mutation rate to estimate past population sizes in whales, researchers found that Pacific gray whales, *Eschrichtius robustus*, were three to five times more abundant before whaling than they are now. And these whales affected sediment resuspension and food webs far more than today's smaller populations.

Trade records show Caribbean marine turtles have declined precipitously. Nesting sites have disappeared and green turtles, which numbered a staggering 15 to 116 million when Columbus arrived, number only 100 nesting females today.

Dusty sales records, fishery yearbooks, and other written material showed HMAP that bluefin tuna arrived in northern European waters by the thousands each summer until an industrialized fishery geared up in the 1920s and literally filled the floors of European fish markets (Figure 9.4). Before World War I, bluefin were rarely caught, but with advanced technology, fishermen in the 1920s were catching 50- to 100-kilogram tuna, and some as big as 700 kilograms. One sports fisherman landed 62 bluefin near the island of Anholt near Denmark in 1928. Abundant tuna in Danish and Swedish waters inspired the formation of the Scandinavian Fishing Club that held bluefin tournaments until the early 1960s. By 1949, fishing was removing 5,500 tons, but by the 1960s catches had plummeted, and bluefin in the North Sea was commercially extinct.

Photos of the winner of Florida fishing contests (Figure 9.5) showed that in the 1950s, large groupers and sharks almost 2 meters in length and weighing 20 kilograms were the prize winners. By the early 1980s fish size had decreased noticeably, and in 2007 the contest winners were small snappers averaging 34 centimeters in length and 2 kilograms in weight.

Hammerhead, thresher, and white sharks have declined in the Northwest Atlantic more than 75% in the last 15 years, and other sharks by more than 50%. Oceanic whitetip and silky sharks, which were the most common sharks in the Gulf of Mexico until the 1950s, have declined more than 99% and 90%, respectively. In the Mediterranean Sea, sharks have declined 99.99% from historical abundances in the early nineteenth to mid-twentieth centuries. These declining numbers illustrate a shifting baseline that leave few people realizing these sharks were once prevalent. Sharks, skates, and rays, like many deep-sea species that grow slowly, reproduce late in life, and produce relatively few offspring, cannot be fished sustainably.

Forecasting change is uncertain and controversial. A projected 100% collapse of global fish stocks by 2048 drew harsh criticism. Reanalysis did find that although about 63% of global fish stocks required rebuilding, some levels of exploitation have moderated to sustainable levels and populations will survive as long as critical thresholds are not exceeded. Still, these analyses were based on the best-known fisheries in the world, with the best experts possible, and excluded lesser-known areas like the tropical Pacific that some believe are fished excessively and unsustainably.

The cascading effects of fishing

Fishing effects reach far and wide, well beyond a single species. Industrial fishing alters food webs by removing top *apex* predators. When trawl gear scrapes the seafloor, much like logging clear cuts a forest, it destroys cold-water corals and other living and non-living bottom habitat. Different fishing gears vary in how selectively they catch fish; some gears remove just about everything they encounter, whereas others target more precisely. Catching unintended species, known as *bycatch*, when fishing for other species changes

Figure 9.5 Diminishing trophies.
Changes in the size and species of trophy fish caught on Key West illustrate that humans have altered marine life in only a few decades. The diminution of the trophy fish on charter boats from **(a)** 1957 to **(b)** early 1980s to **(c)** 2007 dramatizes the loss.

the ocean. Sea turtles, whales, sharks, and seafloor invertebrates become bycatch in shallow- and deep-water fisheries. Modifying fishing gear increases selectivity and decreases bycatch but imperfectly, so the problem is ongoing.

The effects of whole-scale removal of apex predators can cascade through the entire food web, altering species that are not directly related to fishing. Precipitous declines in all 11 species of great sharks in the Northwest Atlantic ecosystem allowed increases of cownose rays that prey on scallops and may have caused the demise of a century-old scallop fishery. In the Pacific, declines of 21% in numbers and 50% in size of sharks and tuna since the 1970s coincided with increases of a few small species that were insufficient to replace the lost biomass. In the Gulf of Mexico, removal of large sharks meant other smaller sharks had no predators and subsequently increased. Between the 1970s and the 1990s in the Mediterranean Sea near Spain and Greece some of the highest steps from the base to the top of the food web disappeared. Increased vulnerability to extinction followed, and biomass at lower trophic levels grew. Fishing, exacerbated by bycatch and habitat degradation, has reduced species richness of coral reef fishes. Increased runoff from land reduces water clarity, increasing algal growth and decreasing coral cover, and all reduce fish habitat and add to fishing impacts. These patterns can be generalized as degraded ecosystems.

Sometimes collapse of one species lets another flourish. After Atlantic cod, *Gadus morhua*, in Newfoundland collapsed, the abundance of one of its prey, northern shrimp, *Pandalus borealis*, exploded. Some fishing jobs were saved as a new industry chased northern shrimp. But fisheries collapses usually force fishermen to chase ever smaller fishes of ever lower economic value that must be harvested in higher numbers to yield profit.

Ocean change beyond fishing

Before sewage treatment, one can imagine foul-smelling streets and canals in Shakespeare's London or Rembrandt's Amsterdam, and pockets of degraded coastline near large cities that date back many centuries. But the scale of coastal hypoxia from fertilizer runoff and sewage input has only been recognized in recent decades. Increases in atmospheric carbon following the second wave of the industrial revolution in the late 1800s and then in the 1950s are linked to slow and steady ocean warming since the 1950s. Only in the last decade have researchers linked this input to ocean acidification.

They have even hypothesized that deep-sea environmental change is linked to warming surface waters and ocean acidity.

Hungry nations now look to *mariculture* – growing cod, salmon, and others – to replace collapsed wild stocks, but at a cost. Mangroves are destroyed to make way for shrimp and other mariculture. Mariculture feeds the high protein diet needed by carnivorous fish by harvesting large abundances of low value ocean life – a strategy likened to trapping shrews and foxes to feed farmed wolves. Other species like mussels don't require food supplement beyond natural sources, but habitat degradation and shoreline change still take a toll.

New analysis shows combined threats are collectively more damaging than the total of their individual effects, and multiple threats simultaneously affect some 41% of the global ocean. Experiments that simulated exploitation and habitat fragmentation demonstrated that populations decline 50 times faster when threats act together than separately. Although the overwhelming drivers of change are exploitation of animals and habitat destruction, adding invasive species and altered climate to the mix may speed change in coming decades.

For millennia the ocean has provided seafood, oils, furs and feathers, and medicines. It regulates climate, gas exchange and oxygen production, and nutrient cycling, and transports people and goods around the world. We relax on beaches and boats, which inspired our artists from Coleridge to Conrad to Hokusai to Doubilet to create beauty.

While the ocean worries us today, we can find hope in its beauty, which persists despite scars. Many fisheries flourish and continue to provide protein to many. Mariculture helps, sometimes with little effect on the ocean when herbivores are cultured. We understand our footprint more clearly than in the past, and many work to sustain ocean biodiversity and abundance into the future. People care about the ocean – even those who live far from its shores. Beyond 2010, humanity can learn more and manage better the changing oceans.

BIBLIOGRAPHY

Alter, S. E., E. Rynes, and S. R. Palumbi, DNA evidence for historic population size and past ecosystem impacts of gray whales. *Proc. Natl. Acad. Sci. USA*, **104**:38 (2007), 15162–7.
Anderson, S. C., H. K. Lotze, and N. L. Shackell, Evaluating the knowledge base for expanding low-trophic-level fisheries in Atlantic Canada. *Can. J. Fish. Aquat. Sci.*, **65**:12 (2008), 2553–71.

Antczak, M. M. and A. Antczak, Between food and symbol: the role of marine molluscs in the late pre-hispanic North-Central Venezuela. In A. Antczak and R. Cipriani (eds.), *Early Human Impact of Megamolluscs British Archaeological Reports S1865* (2008).

Auster, P. J., R. J. Malatesta, R. W. Langton, *et al.*, The impacts of mobile fishing gear on seafloor habitats in the Gulf of Maine (Northwest Atlantic): implications for conservation of fish populations. *Rev. Fish. Sci.*, **4**:2 (1996), 185–202.

Bager, M., M. K. Sondergaard, and B. R. MacKenzie, The development of fisheries at Bornholm, Denmark (Baltic Sea) during 1880s–1914. *Fish. Res.*, **87**:2–3 (2007), 146–54.

Bailey, G. and N. Milner, Molluscan archives from European prehistory. In A. Antczak and R. Cipriani (eds.), *Early Human Impact on Megamolluscs* (Oxford: Archaeopress, 2008), pp. 111–34.

Barrett, J., C. Johnstone, J. Harland, *et al.*, Detecting the medieval cod trade: a new method and first results. *J. Archaeol. Sci.*, **35**:4 (2008), 850–61.

Barrett, J. H., A. M. Locker, and C. M. Roberts, The origins of intensive marine fishing in medieval Europe: the English evidence. *Proc. R. Soc. Lond. B Biol. Sci.*, **271**:1556 (2004), 2417–21.

Barrowman, N. J., R. A. Myers, R. Hilborn, D. G. Kehler, and C. A. Field, The variability among populations of coho salmon in the maximum reproductive rate and depensation. *Ecol. Appl.*, **13**:3 (2003), 784–93.

Baum, J. K., J. M. McPherson, and R. A. Myers, Farming need not replace fishing if stocks are rebuilt. *Nature*, **437**:7055 (2005), 26.

Baum, J. K. and R. A. Myers, Shifting baselines and the decline of pelagic sharks in the Gulf of Mexico. *Ecol. Lett.*, **7**:2 (2004), 135–45.

Baum, J. K., R. A. Myers, D. G. Kehler, *et al.*, Collapse and conservation of shark populations in the Northwest Atlantic. *Science*, **299**:5605 (2003), 389–92.

Baumgartner, T. R., A. Soutar, and V. Ferreirabartrina, Reconstruction of the history of Pacific sardine and northern anchovy populations over the past 2 millennia from sediments of the Santa-Barbara Basin, California. *Calif. Coop. Oceanic Fish. Investig. Rep.*, **33** (1992), 24–40.

Bekker-Nielsen, T., The technology and productivity of ancient sea fishing. In T. Bekker-Nielsen (ed.), *Ancient Fishing and Fish Processing in the Black Sea Region* (Denmark: Aarhus University Press, 2005), pp. 83–95.

Beverton, R. J. H. and S. J. Holt, *On the Dynamics of Exploited Fish Populations* (Ministry of Agriculture, Fisheries and Food, 1957).

Botsford, L. W., J. C. Castilla, and C. H. Peterson, The management of fisheries and marine ecosystems. *Science*, **277**:5325 (1997), 509–15.

Cipriani, R., A. Antczak, and M. Antczak, The study of ancient human-mollusc interactions as an interdisciplinary challenge. In A. Antczak and R. Cipriani (eds.), *Early Human Impact on Megamolluscs* (Oxford: Archaeopress, 2008), pp. 247–54.

Clark, M. R. and J. A. Koslow, Impacts of fisheries on seamounts. In T. Pitcher, T. Morato, P. J. B. Hart, M. Clark, N. Haggan, and R. S. Santos (eds.), *Seamounts: Ecology, Fisheries and Conservation* (Oxford: Wiley-Blackwell, 2007), pp. 413–41.

Clottes, J. and J. Courtin, Neptune's Ice Age gallery. *Nat. Hist.*, **10** (1993).

Coll, M., H. K. Lotze, and T. N. Romanuk, Structural degradation in Mediterranean Sea food webs: testing ecological hypotheses using stochastic and mass-balance modelling. *Ecosystems*, **11**:6 (2008), 939–60.

Cooper, A. B., A. A. Rosenberg, G. Stefansson, and M. Mangel, Examining the importance of consistency in multi-vessel trawl survey design based on the US west coast groundfish bottom trawl survey. *Fish. Res.*, **70**:2–3 (2004), 239–50.

Curtis, R. I., Sources for production and trade of Greek and Roman processed fish. In T. Bekker-Nielsen (ed.), *Ancient Fishing and Fish Processing in the Black Sea Region* (Denmark: Aarhus University Press, 2005), pp. 31–46.

Cushing, D. H., *The Provident Sea* (Cambridge: Cambridge University Press, 1988).

Dayton, P. K., S. F. Thrush, M. T. Agardy, and R. J. Hofman, Environmental effects of marine fishing. *Aquat. Conserv. Mar. Freshwat. Ecosyst.*, **5**:3 (1995), 205–32.

Devine, J. A., K. D. Baker, and R. L. Haedrich, Fisheries: deep-sea fishes qualify as endangered. *Nature*, **439**:7072 (2006), 29.

Díaz, R. J. and R. Rosenberg, Spreading dead zones and consequences for marine ecosystems. *Science*, **321**:5891 (2008), 926–9.

Dulvy, N. K., Y. Sadovy, and J. D. Reynolds, Extinction vulnerability in marine populations. *Fish Fish.*, **4**:1 (2003), 25–64.

Ejstrud, B., Size matters: estimating trade of wine, oil and fish-sauce from Amphorae in the First Century AD. In T. Bekker-Nielsen (ed.), *Ancient Fishing and Fish Processing in the Black Sea Region* (Denmark: Aarhus University Press, 2005), pp. 171–81.

Enghoff, I. B., B. R. MacKenzie, and E. E. Nielsen, The Danish fish fauna during the warm Atlantic period (ca. 7000–3900 BC): forerunner of future changes? *Fish. Res.*, **87**:2–3 (2007), 167–80.

Erlandson, J. M., The archaeology of aquatic adaptations: paradigms for a new millennium. *J. Arch. Res.*, **9**:4 (2001), 287–350.

Evans, P. G. H., *The Natural History of Whales and Dolphins* (New York: Facts on File Publ., 1987).

Feely, R. A., C. L. Sabine, K. Lee, *et al.*, Impact of anthropogenic CO_2 on the $CaCO_3$ system in the oceans. *Science*, **305**:5682 (2004), 362–6.

Ferretti, F., R. A. Myers, F. Serena, and H. K. Lotze, Loss of large predatory sharks from the Mediterranean Sea. *Conserv. Biol.*, **22**:4 (2008), 952–64.

Gage, J. D., J. M. Roberts, J. P. Hartley, and J. D. Humphery, Potential impacts of deep-sea trawling on the benthic ecosystem along the Northern European continental margin: a review. *Am. Fish Soc. Symp.*, **41** (2005), 503–17.

Gallant, T. W., A fisherman's tale. Belgian Archaeological Mission in Greece in collaboration with the Seminar for Greek Archaeology of the State University of Gent, Gent (1985).

Garcia, V. B., L. O. Lucifora, and R. A. Myers, The importance of habitat and life history to extinction risk in sharks, skates, rays and chimaeras. *Proc. R. Soc. Lond. B Biol. Sci.*, **275**:1630 (2008), 83–9.

Gaumiga, R., G. Karlsons, D. Uzars, and H. Ojaveer, Gulf of Riga (Baltic Sea) fisheries in the late 17th century. *Fish. Res.*, **87**:2–3 (2007), 120–5.

Gavriljuk, N. A., Fishery in the life of the nomadic population of the Northern Black Sea area in the Early Iron Age. In T. Bekker-Nielsen (ed.), *Ancient Fishing and Fish Processing in the Black Sea Region* (Denmark: Aarhus University Press, 2005), pp. 105–13.

Gertwagen, R., Approccio multidisciplinare allo studio dell'ambiente marino e della pesca nel Medio Evo nel Mediterraneo orientale. Il Mare Come Era, Proceedings of the II HMAP Mediterranean and the Black Sea project, 27–29 September 2006, Italy (Chioggia, 2008).

Gibson, A. J. F. and R. A. Myers, A meta-analysis of the habitat carrying capacity and maximum reproductive rate of anadromous alewife in Eastern North America. In K. E. Limburg and J. R. Waldman (eds.), *Biodiversity, Status, and Conservation of the World's Shads*. 35th edn (Bethesda, MD, USA: American Fisheries Society, 2003), pp. 211–21.

Guinotte, J. M. and V. J. Fabry, Ocean acidification and its potential effects on marine ecosystems. *Ann. NY Acad. Sci.*, **1134** (2008), 320–42.

Halpern, B. S., S. Walbridge, K. A. Selkoe, *et al.*, A global map of human impact on marine ecosystems. *Science*, **319**:5865 (2008), 948–52.

Harley, C. D. G., A. R. Hughes, K. M. Hultgren, *et al.*, The impacts of climate change in coastal marine systems. *Ecol. Lett.*, **9**:2 (2006), 228–41.

Heithaus, M. R., A. Frid, A. J. Wirsing, and B. Worm, Predicting ecological consequences of marine top predator declines. *Trends Ecol. Evol.*, **23**:4 (2008), 202–10.

Hilborn, R., Comment on "Impacts of biodiversity loss on ocean ecosystem services." *Science*, **316**:5829 (2007), 1281–2.

Holm, P., *Fiskeriets økonomiske betydning i Danmark 1350–1650* (Esbjerg: Sjæk'len, 1999).

Holm, P., History of marine animal populations: a global research program of the Census of Marine Life. *Oceanol. Acta*, **25**:5 (2003), 207–11.

Holm, P., Fishing. In S. I. Krech, J. R. McNeill, and C. Merchant (eds.), *Encyclopedia of World Environmental History* (2004), pp. 529–34.

Holm, P., A. H. Marboe, B. Poulsen, and B. R. MacKenzie, Marine animal populations: a new look back in time. In A. D. McIntyre (ed.), *Life in the World's Oceans: Diversity, Distribution, and Abundance* (Oxford: Blackwell Publishing Ltd., 2010, pp. 3–23.

Hutchings, J. A., Collapse and recovery of marine fishes. *Nature*, **406**:6798 (2000), 882–5.

Jackson, J. B. C., Reefs since Columbus. *Coral Reefs*, **16** (1997), S23–S32.

Jackson, J. B. C., What was natural in the coastal oceans? *Proc. Natl. Acad. Sci. USA*, **98**:10 (2001), 5411–8.

Jacobsen, A. L. L., The reliability of fishing statistics as a source for catches and fish stocks in antiquity. In T. Bekker-Nielsen (ed.), *Ancient Fishing and Fish Processing in the Black Sea Region* (Denmark: Aarhus University Press, 2005), pp. 97–104.

James, M. C., C. A. Ottensmeyer, and R. A. Myers, Identification of high-use habitat and threats to leatherback sea turtles in northern waters: new directions for conservation. *Ecol. Lett.*, **8**:2 (2005), 195–201.

Jenkins, S., M. Burrows, D. Garbary, *et al.*, Comparisons of the ecology of shores across the North Atlantic: do differences in players matter for process? *Ecology*, **89** (2008), S3–S23.

Jones, G. A., Quite the choicest protein dish: the cost of consuming seafood in American restaraunts, 1850–2006. In D. J. Starkey, P. Holm, and M. Barnard (eds.), *Oceans Past: Management Insights from the History of Marine Animal Populations* (London: Earthscan, 2008).

Josephson, E. A., T. D. Smith, and R. R. Reeves, Depletion within a decade: the American 19th century North Pacific right whale fishery. In D. J. Starkey, P. Holm, and M. Barnard (eds.), *Oceans Past: Management Insights from the History of Marine Animal Populations* (London: Earthscan Publications, 2008), pp. 133–49.

Josephson, E., T. D. Smith, and R. R. Reeves, Historical distribution of right whales in the North Pacific. *Fish Fish.*, **9**:2 (2008), 155–68.

Kennedy, V. S., The ecological role of the eastern oyster, *Crassostrea virginica*, with remarks on disease. *J. Shellfish Res.*, **15**:1 (1996), 177–83.

Kher, U., Oceans of nothing, *Time Magazine*, 5 November (2006).

Klaer, N. L., Abundance indices for main commercial fish species caught by trawl from the south-eastern Australian continental shelf from 1918 to 1957. *Mar. Freshw. Res.*, **55**:6 (2004), 561–71.

Klaer, N. L., Changes in the structure of demersal fish communities of the South East Australian Continental Shelf from 1915 to 1961. PhD thesis (2005).

Knowlton, N. and J. B. C. Jackson, Shifting baselines, local impacts, and global change on coral reefs. *PLoS Biol.*, **6**:2 (2008), e54, doi: 10.1371/journal.pbio.0060054.

Kurlansky, M., *Cod: A Biography of the Fish That Changed the World* (New York: Walker & Company, 1998).

Lajus, D. L., J. A. Lajus, Z. V. Dmitrieva, A. V. Kraikovski, and D. A. Alexandrov, The use of historical catch data to trace the influence of climate on fish populations: examples from the White and Barents Sea fisheries in the 17th and 18th centuries. *ICES J. Mar. Sci.*, **62**:7 (2005), 1426–35.

Lawler, A., Archaeology – report of oldest boat hints at early trade routes. *Science*, **296**:5574 (2002), 1791–2.

Levin, L. A., D. F. Boesch, A. Covich, *et al.*, The function of marine critical transition zones and the importance of sediment biodiversity. *Ecosystems*, **4**:5 (2001), 430–51.

Levitus, S., J. I. Antonov, J. Wang, *et al.*, Anthropogenic warming of earth's climate system. *Science,* **292**:5515 (2001), 267–70.

Lotze, H. K., Repetitive history of resource depletion and mismanagement: the need for a shift in perspective. *Mar. Ecol. Prog. Ser.,* **274** (2004), 282–5.

Lotze, H. K., Rise and fall of fishing and marine resource use in the Wadden Sea, southern North Sea. *Fish. Res.,* **87**:2–3 (2007), 208–18.

Lotze, H. K. and M. Glaser, Ecosystem services of semi-enclosed marine systems. In E. R. Urban, B. Sundby, P. Malanotte-Rizzoli, and J. M. Melillo (eds.), *Watersheds, Bays and Bounded Seas* (Washington, DC: Island Press, 2008), pp. 227–49.

Lotze, H. K., H. S. Lenihan, B. J. Bourque, *et al.*, Depletion, degradation, and recovery potential of estuaries and coastal seas. *Science,* **312**:5781 (2006), 1806–9.

Lotze, H. K. and I. Milewski, Two centuries of multiple human impacts and successive changes in a North Atlantic food web. *Ecol. Appl.,* **14**:5 (2004), 1428–47.

Lotze, H. K., K. Reise, B. Worm, *et al.*, Human transformations of the Wadden Sea ecosystem through time: a synthesis. *Helgol. Mar. Res.,* **59**:1 (2005), 84–95.

Lotze, H. K. and B. Worm, Historical baselines for large marine animals. *Trends Ecol. Evol.,* **24**:5 (2009), 254–62.

MacKenzie, B. R., J. Alheit, D. J. Conley, P. Holm, and C. C. Kinze, Ecological hypotheses for a historical reconstruction of upper trophic level biomass in the Baltic Sea and Skagerrak. *Can. J. Fish. Aquat. Sci.,* **59**:1 (2002), 173–90.

MacKenzie, B. R., M. Bager, H. Ojaveer, *et al.*, Multi-decadal scale variability in the eastern Baltic cod fishery 1550–1860 – evidence and causes. *Fish. Res.,* **87**:2–3 (2007), 106–19.

MacKenzie, B. R. and R. A. Myers, The development of the northern European fishery for north Atlantic bluefin tuna *Thunnus thynnus* during 1900–1950. *Fish. Res.,* **87**:2–3 (2007), 229–39.

Marra, J., When will we tame the oceans? *Nature,* **436**:7048 (2005), 175–6.

McClenachan, L., Documenting loss of large trophy fish from the Florida Keys with historical photographs. *Conserv. Biol.,* **23**:3 (2009), 636–43.

McClenachan, L., J. B. C. Jackson, and M. J. H. Newman, Conservation implications of historic sea turtle nesting beach loss. *Front. Ecol. Environ.,* **4**:6 (2006), 290–6.

Mead, J. G. and E. D. Mitchell, The gray whale, *Eschrichtius robustus.* In M. L. Jones, S. L. Swatz, and S. Leatherwood (eds.), *Atlantic Gray Whales* (New York: Academic, 1984), pp. 225–78.

Meldgaard, M., The great auk, *Pinguinus impennis* (L) in Greenland. *Hist. Biol.,* **1** (1988), 145–78.

Mora, C., R. Metzger, A. Rollo, and R. A. Myers, Experimental simulations about the effects of overexploitation and habitat fragmentation on populations facing environmental warming. *Proc. R. Soc. Lond. B Biol. Sci.,* **274**:1613 (2007), 1023–8.

Murawski, S., R. Methot, and G. Tromble, Comment on "Impacts of biodiversity loss on ocean ecosystem services." *Science,* **316**:5829 (2007), 1281.

Myers, R. A., J. K. Baum, T. D. Shepherd, S. P. Powers, and C. H. Peterson, Cascading effects of the loss of apex predatory sharks from a coastal ocean. *Science,* **315**:5820 (2007), 1846–50.

Myers, R. A., S. A. Boudreau, R. D. Kenney, *et al.*, Saving endangered whales at no cost. *Curr. Biol.,* **17**:1 (2007), R10–R1.

Myers, N. and J. Kent, *Perverse Subsidies: How Tax Dollars Can Undercut the Environment and the Economy* (Washington, DC: Island Press, 2001).

Myers, R. A., G. Mertz, and P. S. Fowlow, Maximum population growth rates and recovery times for Atlantic cod, *Gadus morhua. Fish. Bull.,* **95**:4 (1997), 762–72.

Myers, R. A. and B. Worm, Rapid worldwide depletion of predatory fish communities. *Nature,* **423**:6937 (2003), 280–3.

NOAA, Ocean. 2010. Available from: http://www.noaa.gov/ocean.html.

Ojaveer, H. and B. R. MacKenzie, Historical development of fisheries in northern Europe – reconstructing chronology of interactions between nature and man. *Fish. Res.*, **87**:2–3 (2007), 102–5.

Pauly, D., Anecdotes and the shifting baseline syndrome of fisheries. *Trends Ecol. Evol.*, **10**:10 (1995), 430.

Pauly, D., V. Christensen, J. Dalsgaard, R. Froese, and F. Torres, Fishing down marine food webs. *Science*, **279**:5352 (1998), 860–3.

Poulsen, B., *Dutch Herring: An Environmental History, c. 1600–1860* (Amsterdam: Aksant, 2008).

Poulsen, B., P. Holm, and B. R. MacKenzie, A long-term (1667–1860) perspective on impacts of fishing and environmental variability on fisheries for herring, eel, and whitefish in the Limfjord, Denmark. *Fish. Res.*, **87**:2–3 (2007), 181–95.

Probert, P. K., D. G. McKnight, and S. L. Grove, Benthic invertebrate bycatch from a deep-water trawl fishery, Chatham Rise, New Zealand. *Aquat. Conserv. Mar. Freshwat. Ecosyst.*, **7**:1 (1997), 27–40.

Ricker, W. E., Stock and recruitment. *J. Fish. Res. Board Can.*, **11** (1954), 559–623.

Roberts, J. M., A. J. Wheeler, and A. Freiwald, Reefs of the deep: the biology and geology of cold-water coral ecosystems. *Science*, **312**:5773 (2006), 543–7.

Rosenberg, A. A., W. J. Bolster, K. E. Alexander, *et al.*, The history of ocean resources: modeling cod biomass using historical records. *Front. Ecol. Environ.*, **3**:2 (2005), 84–90.

Safina, C., A. A. Rosenberg, R. A. Myers, T. J. Quinn, and J. S. Collie, US ocean fish recovery: staying the course. *Science*, **309**:5735 (2005), 707–8.

Schrope, M., Oceanography – the real sea change. *Nature*, **443**:7112 (2006), 622–4.

Shepherd, T. D. and R. A. Myers, Direct and indirect fishery effects on small coastal elasmobranchs in the northern Gulf of Mexico. *Ecol. Lett.*, **8**:10 (2005), 1095–104.

Sims, D. W. and A. J. Southward, Dwindling fish numbers already of concern in 1883. *Nature*, **439**:7077 (2006), 660.

Smith, K. L., H. A. Ruhl, B. J. Bett, *et al.*, Climate, carbon cycling, and deep-ocean ecosystems. *Proc. Natl. Acad. Sci. USA*, **106**:46 (2009), 19211–8.

Snelgrove, P. V. R., T. H. Blackburn, P. A. Hutchings, *et al.*, The importance of marine sediment biodiversity in ecosystem processes. *Ambio*, **26**:8 (1997), 578–83.

Stolba, V. F., Fish and money: nunismatic evidence for Black Sea fishing. In T. Bekker-Nielsen (ed.), *Ancient Fishing and Fish Processing in the Black Sea Region* (Denmark: Aarhus University Press, 2005), pp. 115–32.

Sumaila, U. R., L. Teh, R. Watson, P. Tyedmers, and D. Pauly, Fuel price increase, subsidies, overcapacity, and resource sustainability. *ICES J. Mar. Sci.*, **65**:6 (2008), 832–40.

Taggart, C., T. J. Anderson, C. Bishop, *et al.*, Overview of cod stocks, biology, and environment in the Northwest Atlantic region of Newfoundland, with emphasis on northern cod. *ICES J. Mar. Sci. Symp.*, **198** (1994), 140–57.

Tittensor, D. P., F. Micheli, M. Nystrom, and B. Worm, Human impacts on the species-area relationship in reef fish assemblages. *Ecol. Lett.*, **10** (2007), 760–72.

TOPP, Tagging of Pacific Predators. 2010. Available from: http://www.topp.org/species/bluefin_tuna.

Trakadas, A., The archaeological evidence for fish processing in the Western Mediterranean. In T. Bekker-Nielsen (ed.), *Ancient Fishing and Fish Processing in the Black Sea Region* (Denmark: Aarhus University Press, 2005), pp. 47–82.

Walsh, B., Can the world's fisheries survive our appetites? *Time Magazine*, 1 August (2009).

Walters, C., Folly and fantasy in the analysis of spatial catch rate data. *Can. J. Fish. Aquat. Sci.*, **60**:12 (2003), 1433–6.

Ward, P. and R. A. Myers, Shifts in open-ocean fish communities coinciding with the commencement of commercial fishing. *Ecology*, **86**:4 (2005), 835–47.

Watling, L. and E. A. Norse, Disturbance of the seabed by mobile fishing gear: a comparison to forest clearcutting. *Conserv. Biol.*, **12**:6 (1998), 1180–97.

Wilberg, M. J. and T. J. Miller, Comment on "Impacts of biodiversity loss on ocean ecosystem services." *Science*, **316**:5829 (2007), 1282.

Wilkins, J., Fish as a source of food in Antiquity. In T. Bekker-Nielsen (ed.), *Ancient Fishing and Fish Processing in the Black Sea Region* (Denmark: Aarhus University Press, 2005), pp. 21–30.

Worm, B., E. B. Barbier, N. Beaumont, *et al.*, Impacts of biodiversity loss on ocean ecosystem services. *Science*, **314**:5800 (2006), 787–90.

Worm, B., R. Hilborn, J. K. Baum, *et al.*, Rebuilding global fisheries. *Science*, **325**:5940 (2009), 578–85.

Worm, B. and H. K. Lotze, Changes in marine biodiversity as an indicator of climate. In T. Letcher (ed.), *Climate Change: Observed Impacts on Planet Earth* (Amsterdam: Elsevier, 2009), pp. 263–79.

Worm, B., H. K. Lotze, I. Jonsen and C. Muir, The future of marine animal populations. In A. D. Mclntyre (ed.), *Life in The World's Oceans Diversity, Distribution, and Abundance* (Oxford: Blackwell Publishing Ltd., 2010), pp. 315–30.

Worm, B. and R. A. Myers, Meta-analysis of cod-shrimp interactions reveals top-down control in oceanic food webs. *Ecology*, **84**:1 (2003), 162–73.

Worm, B., M. Sandow, A. Oschlies, H. K. Lotze, and R. A. Myers, Global patterns of predator diversity in the open oceans. *Science*, **309**:5739 (2005), 1365–9.

PART THREE

From unknown to unknowable

10

Planet Ocean beyond 2010

Accomplishments, many by the Census during the last decade, foreshadow how the goalposts of the known, unknown, and unknowable will shift beyond 2010. Knowledge was extended back in time, providing new baselines of the ocean past that once seemed unknowable. Tools that made the ocean more transparent added knowledge of animal behavior, and new images of the diversity of ocean life. Electronic tags hitchhiked on animals around the ocean deep and shallow, far and wide, transmitting views of Planet Ocean as animal life sees them. Newly discovered species, habitats, and patterns changed our knowledge of the present Planet Ocean. Models project glimpses of future oceans.

Beyond 2010, scientists will explore changed oceans. Census leaders imagined the ocean in 2020 and beyond, and summarized their expectations in Table 10.1. The news is a mixture of the bad, easily predicted from past changes, with some good, arising from better information, communication, and their application to ocean problems. Scientists will acquire more information, the subject in the final two boxes of the table. What does the Census' decade of discovery promise for a more transparent ocean beyond 2010?

Recent additions to known diversity

Some 500^+ research expeditions and countless nearshore samplings discovered thousands of new species from everywhere explored, spanning all the ocean realms. Exploration extended from the Arctic to the Antarctic,

Table 10.1 Predictions of ocean status in 2020 and beyond

Issue	Prediction
More crowded	
Energy extraction	Drilling deeper onto slopes and further north, more platforms, subsea networks, gas hydrate extraction, wind farms, tidal power, wave energy facilities
Ocean transport	More consumer goods, more oil and more ore, more ocean "highways," more Indo-Pacific build-up, new Arctic shipping passage, more cruise ships
Population increase	More people, more people on coasts
Communication	More subsea cables for data, power? More Indo-Pacific demand and build-up
More environmental changes	
Coastal concerns	More storms, sea level rise, more nutrients and hypoxia, more pollutants, less ice, changing sediment supply, reduced freshwater supply, more shoreline modification, more lights
Global concerns	More noise, increasingly acidic, warmer ocean, more stratified, more debris, more spills, bigger fishing vessels, better fish finders and bigger nets
More biological changes	
Rapid biological change	Fewer wild fish, more aquaculture, more "underutilized" species fished, more modified food webs, more diversity changes, more distribution shifts, more extinctions and extirpations, less genetic diversity, more variability of systems, more alien and invasive species, more altered nutrient and carbon cycles, more degraded habitats, more altered migrations
More ocean conflicts	
Unresolved boundaries	Higher stakes (e.g. oil, shipping)
Fewer living resources	More "ownership" of migratory species, more fishing in international waters, more illegal fishing, more piracy
More information	
More transparent	More vessel tracking, more animal tracking, more ocean observatories, more webcams, more floating sensors, more unmanned ocean gliders and other vehicles
Better planning	More information flow (Wiki etc.), more data freely available, better forecasting models (weather, ecological), more public awareness, more marine reserves and protected areas with better design, more species information, better collaboration?, better governance?

along shorelines onto coral reefs and into the Gulf of Maine. Exploration extended down the continental slope through seeps and vents into the abyss, and up onto seamounts and along the Mid-Atlantic Ridge. Exploration added knowledge of diversity of new species, their distribution around Planet Ocean, and their abundances. New knowledge was added for the smallest microbes and zooplankton and the largest whales. Some species thought to be extinct are now known to live.

New applications of genetic technology move cryptic and unknown species into the known and knowable, whether to compare microbes in a single sample, to contrast copepod crustaceans from distant locations in the global ocean, or to categorize untold numbers of previously unseen species from coral reefs. Discoveries of the hottest, deepest, northernmost, and southernmost vents, and the largest known cold seep on Earth all add to the riot of species. Rarity is now thought common, and common is rare, with species from microbes to fish seen only once in all of human exploration, pointing to a plethora of life.

Recent additions to known distributions and abundances

While new species were being discovered, scientists have also been adding new knowledge about the distributions and abundances of known species. The OBIS central database reconciled vast and varied data on known ocean life into standardized and comparable form in collaboration with the World Register of Marine Species. OBIS works with WoRMS to put definitive names on all the almost 250,000 known species of marine life. Anyone in the world can access whatever their need almost 28 million records (and counting) of individual ocean specimens now amassed in a single database (http://www. iobis.org). Global datasets and novel analyses populate new maps of ocean life that identify hot spots of diversity and abundance (Figure 10.1), some unexpected. Standardized sampling tools and strategies with taxonomic catalogs facilitate the mapping and enhance less ambiguous and controversial analyses.

Thousands of tagged animals spanning 40 different species show new migration "highways" and "truck stops" far from land within fluid borders that animals "see." Electronic sensors show the tagged animals' view of Planet Ocean as they move to feed and reproduce. They tell us that events on one side of the Pacific affect the other, and that events in the Antarctic link to those in the Arctic and the ocean between. Above the Mid-Atlantic Ridge the tagged animals move up and down as day changes to night. New acoustic imaging of swaths of ocean shows well-organized schools of millions of fish that cover an area the size of Manhattan Island, assembling within minutes.

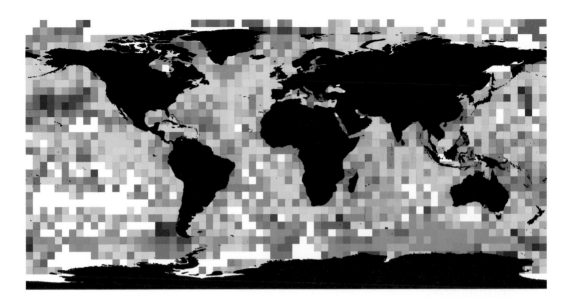

Hurlbert's Index – es(50)

Low   High

Figure 10.1 Hot spots of ocean diversity can now be located.
Global datasets encompassing Planet Ocean now locate the highest diversity.
Hurlbert's index of diversity tells how many different species are found in a random
sample of 50 individuals from 5° x 5° squares in the ocean. In the figure, red indicates
high species diversity, blue low. In the white areas, the dataset did not contain the
required minimum 50 specimens to calculate the index.

Applications of the Census discoveries

The Census shows a new way of conducting science with a global
network of 2,700 scientists tackling complex questions with a suite of tools
from traditional to cutting edge, from simple to complex. The power of one
is the combined strength of many, creating new views and new excitement.
Ocean biodiversity has moved from understudy to mainstage player,
and the Census has led.

Ocean life is freshly visualized with new tools like Google Earth's Ocean
layer and other three-dimensional views that show the pattern and movement
of ocean life. The appeal of marine biota is evident in the popular press

and the findings of the Census, whether bad news on the state of some organisms and ecosystems today, or the breathtaking and inspiring imagery of new species that fascinates the public.

The Census moves fascination with "stamp collecting" on to conservation of diversity, distribution, and abundance. Monitoring Arctic, Antarctic, and coral reef ecosystems closes gaps in knowledge. ArcOD participants work within the international *Arctic Council* to monitor how Arctic marine ecosystems respond to climate change. CAML contributed to the *Commission for the Conservation of Antarctic Marine Living Resources'* decision to declare two Vulnerable Marine Ecosystems and protect them from long-line fishing. A new Census study shows most Marine Protected Areas (MPAs) today don't conserve the reefs they were intended to protect, and they can't offset declines in coral reef ecosystems. Managers have embraced CReefs tools in new plans to monitor acidification and coral reef health beyond 2010.

TOPP tagging data on black-footed albatross were used by the *US Fish and Wildlife Service* during international deliberations on conservation. The establishment of an MPA to protect loggerhead turtles off the coast of Baja California was spurred by TOPP findings, and the discovery of leatherback turtle migration highways has contributed to a resolution by the *International Union for Conservation of Nature* (IUCN) to protect leatherbacks in the open ocean. Knowledge of specific migration routes shows where species are vulnerable and improves conservation of vulnerable species by reducing bycatch and trash. POST data on salmon survival in river dam systems shows that dam removal may not produce expected benefits.

CenSeam contributed to the *Food and Agriculture Organization of the United Nations* (FAO) efforts to develop guidelines for sustainable management of seamount ecosystems and their fisheries. Classification and predictive tools on seamount distributions will aid in future development of seamount MPA strategies. ChEss scientists contributed to plans to protect a network of chemosynthetic environments and scientific guidelines and criteria to be considered by the *Convention on Biological Diversity*. A code of conduct for scientific study of hydrothermal vents has been developed with ChEss input, and is utilized by the *OSPAR Commission*, the lead agency on Northeast Atlantic conservation. CeDAMar has worked with the *International Seabed Authority* to identify the need and strategy to create MPAs within manganese nodule areas of the abyssal plains.

The still unknown

During its decade the Census clarified where gaps in knowledge remain to be filled beyond 2010. Different locations, habitats, seasons, and taxonomic groups provide needs and opportunities for research, some more than others.

Discoveries of unknown species in the deep Southern Ocean seafloor reveal one clearly neglected region ripe for exploration. Our ignorance of life beneath ice-covered regions of the Arctic and Antarctic holds opportunity for new discoveries. Sampling polar ecosystems beneath the ice during winter will produce surprising new discoveries. Many areas of the deep sea, especially the meso- and bathypelagic zones encompassing 200–4,000 meters depth (Figure 7.1), and deep ocean trenches will likely yield discoveries. Only a small fraction of global seamounts have been sampled quantitatively. The discovery of vents and seeps in the Arctic and Southern Ocean suggests more will follow, along with discovery of new species. With more unknowns per square-meter than anywhere in the ocean or possibly on Earth, coral reefs provide opportunities for research and threats give them priority.

A taxonomic inventory of marine life requires a census of microbes, which are far more unknown than known. Small invertebrates, particularly in the deep sea, are little known. Although new species of fish will be encountered, particularly in the deep sea, they will be discovered at a lower rate than will invertebrates.

Biodiversity and management needs

Some ocean biodiversity research needs span from applied to pure research. Little is known about how the riot of species is linked to the ways the ocean functions, and whether species loss will compromise ocean health. This gap links to a need to understand the resilience of the few relatively pristine ecosystems that remain on Planet Ocean – the deep sea, Arctic, and Antarctic. These systems will face increasing pressures from human activities, as will other ocean ecosystems. We know more about where adults and juveniles move in the ocean, but little about how their eggs and larval offspring are transported and how their fate affects abundance and spatial pattern in nature. All of these questions have relevance to MPA design and management of species from fishes to scallops to lobsters. We also need to know which habitats are critical to different species, and where those critical habitats are found.

Abundance, the third of the Census triad, has proved more elusive than diversity and distribution. Knowing whether marine life, its numbers and biomass, is growing or declining rests on reliable measurements of abundance, a clear priority beyond 2010. Surveyors finally have a comprehensive framework into which they can put information of more than 200,000 species.

The unknown, but soon knowable

We will soon become "virtual fish" that experience the entirety of the fish's world through computers in our warm, dry offices. No need to wait for reincarnation! As tracking tools advanced in the last decade, future advances can be foreseen. "Business card" tags now in development will send and receive information, so tagged animals share their data as they pass in the wild, creating swimming data libraries. Tags will soon simultaneously communicate physiology, oceanography, and location. Smaller tags with better batteries will track smaller organisms and stages of their lives, longer and farther. Underwater GPS, if possible, would expand the types of organisms tracked and applications. These sensors will never fully replace research ships, but they will augment them, especially for poorly known and inaccessible waters of Planet Ocean. The Ocean Tracking Network already tracks tagged animals with an emerging global network of listening stations.

We will continue to work toward a barcoding device similar to the tricorder popularized in the television show *Star Trek* that we can place in the ocean to scan and identify passing organisms, perhaps if they shed a little DNA or other clues like amino acids. Environmental genetic sequencing, perhaps in tandem with continuous plankton recorders or acoustic and optical sensors with automated recognition software may form a sensory package to tell us the global diversity, distribution, and abundance of species.

Already satellites continuously orbit the Earth, collecting imagery for myriad applications. Soon, autonomous underwater vehicles (AUVs) may move across the seafloor like underwater satellites. AUVs today run well-defined, pre-programmed missions, but much as cars on our terrestrial highways need fuel, the next-generation AUV may visit underwater docking stations to recharge batteries and download data. Extended missions could map entire regions without human presence, using acoustic imaging of larger and larger areas of seabed. Remotely operated vehicles (ROVs) now have better optics, manipulator arms, and sensors than even a decade ago, and do many tasks economically that submersibles did in the past.

With new predictive analyses and models we will sample the ocean more effectively, placing effort where it will tell us the most. We will extract important information from noisy and complex data with greater certainty in our predictions. We can answer more precisely when asked how much fish can be harvested sustainably, or which ocean regions should be protected. The many global databases from marine microbes to mammals captured in OBIS are an enduring Census legacy. Refined spatial resolution in many physical measurements of the environment is ripe for overlays of biodiversity knowledge that will tell us much about diversity, distribution, and abundance of ocean life. Those overlays have begun, and possibilities are many.

As Census partner programs, the *Encyclopedia of Life* (EOL) and *World Register of Marine Species* (WoRMS) add new species, and as digitization of published work progresses, we will easily confirm names and characteristics of species by clicking through the Internet to its unique page. We will view and download photographs and published studies on that species. A few keystrokes will replace many hours spent in dusty libraries as graduate students. Funding agencies and scientists will demand data sharing and integration. Data rescue will add irreplaceable knowledge and reduce redundant effort. Today with OBIS, we can quickly see if the sea star we found on the shoreline was seen there before, and where it occurs globally. OBIS does not include all of the ocean diversity data ever collected, but as it builds into many millions of records, so does the utility and potential analyses of the database. Birdwatchers have used this strategy for years; the search term "birdwatchers' network" alone produces over a million hits on Google!

The Census has focused on key areas, but not all ocean life. Groups like phytoplankton were the focus of previous international science programs, and therefore not a central focus of the census. Some environments such as deep ocean trenches were little sampled.

Contaminants in marine food chains and in water flowing into the ocean were also outside the purview of the first global Census, but are hugely important in a changing ocean. Carcinogens, heavy metals, human hormones, and other pollutants cause change and require bans on seafood consumption. Even deep-sea ecosystems now contain measurable contaminants from human activities. These topics are particularly compelling issues within our expanded view of ocean biodiversity.

The unknowable

The truly unknowable of ocean biodiversity, at least in the immediate future, is very little, except in terms of absolutes. As someone who has studied marine biodiversity my entire career, I find it hard to admit the numbers and identities of all marine species on Planet Ocean may be unknowable. With the tools now available, our capacity to catalog species will continue to accelerate, assuming of course that we can use these tools to overcome the impediment of too few qualified taxonomists and too many species to identify and name. We will eventually know the identities of most species in most environments, the distributions of many species in many environments, and abundances of common species. But the size of the ocean, the remoteness of many habitats, and the limited global scientific capacity will doom some species to remain unknown, perhaps forever. Some of these unknown species will go extinct without us ever seeing them and others will continue to quietly do whatever it is they do in Planet Ocean.

Epilogue on an ocean Census

The Census affirmed that different people need different knowledge about marine life. Those who develop policy want information for their own backyard and in summary form that clearly illustrates pictures of diversity, distribution, and abundance. Managers want to extract living resources like fish and non-living resources like manganese nodules without compromising ocean health. The public wants to know which seafood is harvested sustainably and which regions are hot spots for ocean life and should be protected. Fishermen ask, "Where are the fish?" They want jobs, and if closures cost jobs then decisions must have a sound basis. In short, new research opportunities can produce knowledge for diverse players in ocean use and management. The players range from conservationists to fisheries managers to international organizations such as the *International Council for the Exploration of the Sea* (ICES), the *North Pacific Marine Science Organization* (PICES), or the *International Commission for the Conservation of Atlantic Tunas* (ICCAT). All players are part of the translation of scientific discovery to on-the-ground application on and above the seafloor.

Knowledge is power, but just as Lankester's counter to Huxley's excessive optimism was overshadowed by the economic needs and politics of that time, new knowledge on ocean life is most powerful when political and societal players act on that knowledge for the long-term benefit of humanity and Planet Ocean. We know much about life in the ocean, and we know a lot of what we do as a hungry society is unsustainable with unpredictable consequences. But change is in the wind. In several well-studied systems, exploitation rates have declined and catch rates have been pushed below a critical threshold that should maintain populations at a healthy and sustainable level. Conservation efforts directed at higher trophic levels have stabilized some populations and allowed others to begin rebuilding. Marine protected areas and closures work in many marine ecosystems, particularly as knowledge gaps that hinder effective design are filled. Still more tough choices must be made as humanity crowds the ocean.

Though conducted as a 10-year program with a fixed endpoint, another major legacy of this decade of discovery is the creation of research links that will continue the work of the Census into the future. Recently launched national programs such as the *Canadian Healthy Oceans Network* (CHONe), or programs in India and Korea, will continue ocean biodiversity work beyond 2010, as will international collaborations between and within Census projects where some field efforts will go on. The Ocean Tracking Network was made possible by the successes of the Census in tracking organisms and the efforts of Census Co-Senior Scientist Ron O'Dor. Few of these international programs would have seen the light of day without the Census. Even without funded field programs, exchange of specimens, data, and other information will continue to flow for at least the current generation of scientists. These programs will work to address questions identified by the Census and other questions unique to a particular region. The Census has emphasized sharing ideas and technologies between partners from countries wealthy and poor. Students have seen a wide range of possibilities and knowledge gaps of opportunity for future study of marine life.

The age of discovery in the ocean has certainly not passed and more is discovered that demands preservation. We must do better. Whether a factory worker in Ontario or yak herder in Tibet has seen the ocean, swam in its spectacular diversity of life, or dined on its edible riches, they need to know that every environment on Earth is tied to marine life. We will all sink

Figure 10.2 Hope for Planet Ocean.
A skin diver quietly accompanies a sunfish, *Mola mola,* swimming through Pacific water. Human and fish illustrate the beauty and coexistence of humanity and marine life. The Census has shown that marine life can be resilient. Where the human affinity for marine life causes its protection, it can recover.

or swim together (Figure 10.2). As I look out at the ocean below my window and type these closing words, I am convinced the enthusiasm for ocean discovery generated by the Census will carry forward far beyond 2010. We have at least a million reasons and opportunities for wonder, excitement, and hope. *Provehito in Altum,* making ocean life count.

BIBLIOGRAPHY

Amaral-Zettler, L., L. F. Artigas, J. Baross, *et al.*, A global census of marine microbes. In A. D. McIntyre (ed.), *Life in the World's Oceans: Diversity, Distribution, and Abundance* (Oxford: Blackwell Publishing Ltd., 2010), pp. 223–45.

Anderson, A., Introducing the transparent ocean. *The Economist* (2009).

Appeltans, W., P. Bouchet, G. A. Boxshell, *et al.*, World Register of Marine Species. 2010. Available from: http://www.marinespecies.org.

Baba, K., E. Macpherson, G. C. B. Poore, *et al.*, Catalogue of squat lobsters of the world (Crustacea: Decapoda: Anomura – families Chirostylidae, Galatheidae and Kiwaidae). *Zootaxa*, **1905** (2008), 3–220.

Baker, M. C., E. Z. Ramirez-Llodra, P. A. Tyler, *et al.*, Biogeography, ecology and vulnerability of chemosynthetic ecosystems in the deep sea. In A. D. McIntyre (ed.), *Life in the World's Oceans: Diversity, Distribution, and Abundance* (Oxford: Blackwell Publishing Ltd., 2010), pp. 161–82.

Barone, J., The 6 most important experiments in the world. *Discover Magazine* (December 2007).

Benedetti-Cecchi, L., I. Bertocci, S. Vaselli, E. Maggi, and F. Bulleri, Neutrality and the response of rare species to environmental variance. *PLoS ONE*, **3**:7 (2008), e2777, doi: 10.1371/journal.pone.0002777.

Birks, J. B. (ed.), *Rutherford at Manchester* (London: Heywood, 1962).

Block, B. A., D. P. Costa, and S. J. Bograd, A view of the ocean from Pacific predators. In A. D. McIntyre (ed.), *Life in the World's Oceans: Diversity, Distribution, and Abundance* (Oxford: Blackwell Publishing Ltd., 2010), pp. 291–311.

Brainard, R., S. Bainbridge, and R. Brinkman, An international network of coral reef observing systems (I-CREOS), CWP-3A-02. In *Proceedings of OceanObs'09: Sustained Ocean Observations and Information for Society* Vol. 1, Venice, Italy, 21–25 September 2009, J. Hall, D. E. Harrison, and D. Stammer, (eds.), ESA Publication WPP-306.

Brandt, A., A. J. Gooday, S. N. Brandao, *et al.*, First insights into the biodiversity and biogeography of the Southern Ocean deep sea. *Nature*, **447**:7142 (2007), 307–11.

Bucklin, A. C., S. Nishida, S. Schnack-Schiel, *et al.*, A Census of zooplankton of the global ocean. In A. D. McIntyre (ed.), *Life in the World's Oceans: Diversity, Distribution, and Abundance* (Oxford: Blackwell Publishing Ltd., 2010), pp. 247–65.

CCAMLR, *Report of the Twenty-seventh Meeting of the Commission. Commission for the Conservation of Antarctic Marine Living Resources* (Hobart, 2008).

Consalvey, M., M. R. Clark, A. A. Rowden, and K. I. Stocks, Life on seamounts. In A. D. McIntyre (ed.), *Life in the World's Oceans: Diversity, Distribution, and Abundance* (Oxford: Blackwell Publishing Ltd., 2010), pp. 123–38.

Costello, M. J., Distinguishing marine habitat classification concepts for ecological data management. *Mar. Ecol. Prog. Ser.*, **397** (2009), 253–68.

Crist, D. T., G. Scowcroft, and J. M. Harding, *World Ocean Census: A Global Survey of Marine Life* (Buffalo, NY: Firefly Books, 2009).

Diaz, J. M., F. Gast, and D. C. Torres, Rediscovery of a Caribbean living fossil: *Pholadomya candida* GB Sowerby I, 1823 (Bivalvia: Anomalodesmata: Pholadomyoidea). *Nautilus*, **123**:1 (2009), 19–20.

Drogin, B., Mapping an ocean of species. *Los Angeles Times* (2 August 2009).

Ebbe, B., D. S. M. Billett, A. Brandt, *et al.*, Diversity of abyssal marine life. In A. D. McIntyre (ed.), *Life in the World's Oceans: Diversity, Distribution, and Abundance* (Oxford: Blackwell Publishing Ltd., 2010), pp. 139–60.

Eicken, H., R. Gradinger, M. Salganek, *et al.*, *Field Techniques for Sea-ice Research* (Fairbanks: University of Alaska Press, 2009).

Feely, R. A., V. J. Fabry, and A. G. Dickson, An international observational network of ocean acidification. In *Proceedings of OceanObs'09: Sustained Ocean Observations and Information for Society* Vol. 1, Venice, Italy, 21–25 September 2009, J. Hall, D. E. Harrison, and D. Stammer (eds.), ESA Publication WPP-306.

Fogarty, M. J. and S. A. Murawski, Do marine protected areas really work? *Oceanus,* **43**:2 (2005), 1–3.

German, C. R., D. R. Yoerger, M. Jakuba, *et al.,* Hydrothermal exploration with the Autonomous Benthic Explorer. *Deep Sea Res. I,* **55**:2 (2008), 203–19.

Giangrande, A., Biodiversity, conservation and "taxonomic impediment." *Aquat. Conserv. Mar. Freshwat. Ecosyst.,* **13** (2003), 451–9.

Gradinger, R., B. A. Bluhm, R. R. Hopcroft, *et al.,* Marine life in the Arctic. In A. D. McIntyre (ed.), *Life in the World's Oceans: Diversity, Distribution, and Abundance* (Oxford: Blackwell Publishing Ltd., 2010), pp. 183–202.

Gutt, J., G. Hosie, and M. Stoddart, Marine life in the Antarctic. In A. D. McIntyre (ed.), *Life in the World's Oceans: Diversity, Distribution, and Abundance* (Oxford: Blackwell Publishing Ltd., 2010), pp. 203–20.

Halpern, B. S., S. Walbridge, K. A. Selkoe, *et al.,* A global map of human impact on marine ecosystems. *Science,* **319**:5865 (2008), 948–52.

Holm, P., A. H. Marboe, B. Poulsen, and B. R. MacKenzie, Marine animal populations: a new look back in time. In A. D. McIntyre (ed.), *Life in the World's Oceans: Diversity, Distribution, and Abundance* (Oxford: Blackwell Publishing Ltd., 2010), pp. 3–23.

Hosie, G., D. M. Stoddart, and V. Wadley, *The Census of Antarctic Marine Life and the Australian-French-Japanese CEAMARC (Collaborative East Antarctic Marine Census) contribution.* International Symposium Asian Collaboration in IPY 2007–2008 Tokyo : National Institute of Polar Research, 2007).

Knowlton, N., *Citizens of the Sea: Wondrous Creatures from the Census of Marine Life* (Florida: National Geographic, 2010).

Knowlton, N., R. E. Brainard, R. Fisher, *et al.,* Coral reef biodiversity. In A. D. McIntyre (ed.), *Life in the World's Oceans: Diversity, Distribution, and Abundance* (Oxford: Blackwell Publishing Ltd., 2010), pp. 65–77.

Leathwick, J., A. Moilanen, M. Francis, *et al.,* Novel methods for the design and implementation of marine protected areas in offshore waters. *Conserv. Lett.,* **1**:2 (2008), 91–102.

Levin, L. A., D. F. Boesch, A. Covich, *et al.,* The function of marine critical transition zones and the importance of sediment biodiversity. *Ecosystems,* **4**:5 (2001), 430–51.

Lotze, H. K., H. S. Lenihan, B. J. Bourque, *et al.,* Depletion, degradation, and recovery potential of estuaries and coastal seas. *Science,* **312**:5781 (2006), 1806–9.

Makris, N. C., P. Ratilal, S. Jagannathan, *et al.,* Critical population density triggers rapid formation of vast oceanic fish shoals. *Science,* **323**:5922 (2009), 1734–7.

Menot, L., M. Sibuet, R. S. Carney, *et al.* New perceptions of continental margin biodiversity. In A. D. McIntyre (ed.), *Life in the World's Oceans: Diversity, Distribution, and Abundance* (Oxford: Blackwell Publishing Ltd., 2010), pp. 79–101.

Mora, C., A clear human footprint in the coral reefs of the Caribbean. *Proc. R. Soc. Lond. B Biol. Sci.,* **275**:1636 (2008), 767–73.

Mora, C., S. Andrefouet, M. J. Costello, *et al.,* Coral reefs and the global network of marine protected areas. *Science,* **312**:5781 (2006), 1750–1.

Mora, C., D. P. Tittensor, and R. A. Myers, The completeness of taxonomic inventories for describing the global diversity and distribution of marine fishes. *Proc. R. Soc. Lond. B Biol. Sci.,* **275**:1631 (2008), 149–55.

OBIS, OBIS: explore data on locations of marine animals and plants. 2010. Available from: http://www.iobis.org.

Payne, J., K. Andrews, C. Chittenden, *et al.*, Tracking fish movements and survival on the Northeast Pacific Shelf. In A. D. McIntyre (ed.), *Life in the World's Oceans: Diversity, Distribution, and Abundance* (Oxford: Blackwell Publishing Ltd., 2010), pp. 269–90.

Peckham, S. H., D. M. Diaz, A. Walli, *et al.*, Small-scale fisheries bycatch jeopardizes endangered Pacific loggerhead turtles. *PLoS ONE*, **2**:10 (2007), e1041.

Rasmussen, K., D. M. Palacios, J. Calambokidis, *et al.*, Southern Hemisphere humpback whales wintering off Central America: insights from water temperature into the longest mammalian migration. *Biol. Lett.*, **3**:3 (2007), 302–5.

Rigby, P. R., K. Iken, and Y. Shirayama, *Sampling Biodiversity in Coastal Communities: NaGISA Protocols for Seagrass and Macroalgal Habitats* (Japan: Kyoto University Press, 2007).

Roberts, C. M., J. A. Bohnsack, F. Gell, J. P. Hawkins, and R. Goodridge, Effects of marine reserves on adjacent fisheries. *Science*, **294**:5548 (2001), 1920–3.

Rogers, A. D., M. R. Clark, J. Hall-Spencer, and K. M. Gjerde, *The Science Behind the Guidelines: A Scientific Guide to the FAO Draft International Guidelines* (Switzerland: IUCN, 2008).

Rowden, A. A., M. R. Clark, and I. C. Wright, Physical characterisation and a biologically focused classification of "seamounts" in the New Zealand region. *NZ J. Mar. Freshwat. Res.*, **39**:5 (2005), 1039–59.

Sale, P. F., R. K. Cowen, B. S. Danilowicz, *et al.*, Critical science gaps impede use of no-take fishery reserves. *Trends Ecol. Evol.*, **20**:2 (2005), 74–80.

Shaffer, S. A., Y. Tremblay, H. Weimerskirch, *et al.*, Migratory shearwaters integrate oceanic resources across the Pacific Ocean in an endless summer. *Proc. Natl. Acad. Sci. USA*, **103**:34 (2006), 12799–802.

Shillinger, G. L., D. M. Palacios, H. Bailey, *et al.*, Persistent leatherback turtle migrations present opportunities for conservation. *PLoS Biol.*, **6**:7 (2008). e171, doi: 10.1371/journal.pbio.0060171.

Sims, D. W. and A. J. Southward, Dwindling fish numbers already of concern in 1883. *Nature*, **439**:7077 (2006), 660.

Smith, C. R., G. L. J. Paterson, P. J. D. Lambshead, *et al.*, *Biodiversity, Species Ranges, and Gene Flow in the Abyssal Pacific Nodule Province: Predicting and Managing the Impacts of Deep Seabed Mining* (Kingston, Jamaica: International Seabed Authority, 2008).

Snelgrove, P. V. R., M. C. Austen, G. Boucher, *et al.*, Linking biodiversity above and below the marine sediment-water interface. *Bioscience*, **50**:12 (2000), 1076–88.

Sobel, D., Stalking fish in the name of science. *Discover Magazine* (September 2009).

Sogin, M. L., H. G. Morrison, J. A. Huber, *et al.*, Microbial diversity in the deep sea and the underexplored "rare biosphere." *Proc. Natl. Acad. Sci. USA*, **103** (2006), 12115–20.

Stokesbury, M. J. W., M. J. Dadswell, K. N. Holland, *et al.*, Tracking of diadromous fishes at sea using hybrid acoustic and archival electronic tags. In R. Cunjak (ed.), *Challenges for Diadromous Fishes in a Dynamic Global Environment* (Bethesda, MD: American Fisheries Society, 2009).

Teo, S. L. H., R. M. Kudela, A. Rais, *et al.*, Estimating chlorophyll profiles from electronic tags deployed on pelagic animals. *Aquat. Biol.*, **5**:2 (2009), 195–207.

Tittensor, D., A. R. Baco, P. E. Brewin, *et al.*, Predicting global habitat suitability for stony corals on seamounts. *J. Biogeogr.*, **36**:6 (2009), 1111–28.

Tittensor, D. P., C. Mora, W. Jetz, *et al.*, Global patterns and predictors of marine biodiversity across taxa. *Nature* (2010) doi: 10.1038/nature09329.

Vanden Berghe, E., K. I. Stocks, and J. F. Grassle, Data integration: the Ocean Biogeographic Information System. In A. D. McIntyre (ed.), *Life in the World's Oceans: Diversity, Distribution, and Abundance* (Oxford: Blackwell Publishing Ltd., 2010), pp. 333–53.

Vecchione, M., O. A. Bergstad, I. Byrkjedal, *et al.*, Biodiversity patterns and processes on the Mid-Atlantic Ridge. In A. D. McIntyre (ed.), *Life in the World's Oceans: Diversity, Distribution, and Abundance* (Oxford: Blackwell Publishing Ltd., 2010), pp. 103–21.

Vermeulen, N., *Supersizing Science: On Building Large-scale Research Projects in Biology* (Maastricht: Maastricht University Press, 2009).

Vongraven, D., P. Arneberg, I. Bysveen, *et al.*, *Circumpolar marine biodiversity monitoring plan.* Background paper. CAFF CBMP Report No. 19 (CAFF International Secretariat, Akureyri, Iceland, 2009).

Walsh, B., Can the world's fisheries survive our appetites? *Time* (2009).

Welch, D. W., E. L. Rechisky, M. C. Melnychuk, *et al.*, Survival of migrating salmon smolts in large rivers with and without dams. *PLoS Biol.*, **6**:10 (2008), e265, doi: 10.1371/journal.pbio.0060265.

Worm, B., R. Hilborn, J. K. Baum, *et al.*, Rebuilding global fisheries. *Science*, **325**:5940 (2009), 578–85.

Worm, B., H. K. Lotze, and R. A. Myers, Predator diversity hotspots in the blue ocean. *Proc. Natl. Acad. Sci. USA*, **100**:17 (2003), 9884–8.

Worm, B., M. Sandow, A. Oschlies, H. K. Lotze, and R. A. Myers, Global patterns of predator diversity in the open oceans. *Science*, **309**:5739 (2005), 1365–9.

Young, L. C., C. Vanderlip, D. C. Duffy, V. Afanasyev, and S. A. Shaffer, Bringing home the trash: do colony-based differences in foraging distribution lead to increased plastic ingestion in Laysan albatrosses? *PLoS ONE*, **4**:10 (2009), e7623, doi: 10.1371/journal.pone.0007623.

FIGURE CREDITS

CHAPTER 1

Figure 1.1   Credit: Jesse Cleary and Ben Donnelly, Census Mapping and Visualization, Marine Geospatial Ecology Lab, Duke University, USA.

Figure 1.2a   Photo: Carola Espinoza, FONDECYT 1070552, Universidad de Concepción, Chile.

Figure 1.2b   Photo: Marsh Youngbluth, MAR-ECO.

Figure 1.2c   Photo: ©Monterey Bay Aquarium Foundation/Randy Wilder, USA.

Figure 1.2d   Photo: Russ Hopcroft, University of Alaska Fairbanks, USA.

Figure 1.2e   Photo: Kacy Moody, 2004, Corpus Christi, Texas, USA.

Figure 1.2f   Photo: MAR-ECO, Tracey Sutton, Virginia Institute of Marine Science, USA.

Figure 1.3   Credit: Ei Fujioka, Census Mapping and Visualization, Marine Geospatial Ecology Lab, Duke University, USA.

Figure 1.4   Credit: Ben Donnelly, Census Mapping and Visualization, Marine Geospatial Ecology Lab, Duke University, USA.

Figure 1.5a   Photo: Alexis Fifis, IFREMER, France.

Figure 1.5b   Photo: Marina R. Cunha, CESAM-Universidade de Aveiro, Portugal. P. C. Dworschak and M. R. Cunha, "A new subfamily, Vulcanocalliacinae n. subfam., for *Vulcanocalliax arutyunovi* n. gen., n. sp. from a mud volcano in the Gulf of Cádiz (Crustacea, Decapoda, Callianassidae)." *Zootaxa*, **1460**(2007) 35–46.

Figure 1.5c   Photo: Max K. Hoberg, University of Alaska Fairbanks, USA.

Figure 1.5d   Photo: Tin-Yam Chan, National Taiwan Ocean University, Keelung. S. T. Ahyong; T.-Y. Chan and P. Bouchet, "Mighty claws: A new genus and species of lobster from the Philippine deep sea" (Crustacea: Decapoda: Nephropidae). *Zoosystema*, 32:3 (in press).

Figure 1.5e   Photo: Russ Hopcroft, University of Alaska Fairbanks, USA.

Figure 1.5f   Photo: Richard Pyle, Bishop Museum, Hawaii, USA.

Figure 1.5g   Photo: Elaina M. Jorgensen, University of Washington, USA.
M. Vecchione, L. Allcock, U. Piatkowski and J. Strugnell,
Benthoctopus rigbyae, n. sp., a new species of Cephalopod
(Octopoda; Incirrata) from near the Antarctic Peninsula.
Malacologia, **51**(2009) 13–28, Fig. 4.

Figure 1.5h   Photo: Cédric d'Udekem d'Acoz, The Royal Belgian Institute of
Natural Sciences, Brussels.

Figure 1.5i   Photo: Charles Griffiths, University of Cape Town, South Africa.

Figure 1.5j   Photo: Peter Rask Moller, MAR-ECO.

CHAPTER 2

Figure 2.1   Photo: Pascal Kobeh, Galatée Films, France.

Figure 2.2   Photo: Pascal Kobeh, Galatée Films, France.

Figure 2.3a   Photo: Elizabeth Calvert Siddon, University of Alaska
Fairbanks, USA.

Figure 2.3b   Photo: Shawn Harper, University of Alaska Fairbanks, USA.

Figure 2.4   Photo: Canadian Scientific Submersible Facility (CSSF)/ROPOS,
Peter Lawton and Anna Metaxas, Canada.

Figure 2.5   Photo: Courtesy of Richard A. Lutz, The Stephen Low Company,
Woods Hole Oceanographic Institution, and Emory Kristof of the
National Geographic Society, USA.

CHAPTER 3

Figure 3.1a   Photo: Chris Linder, Woods Hole Oceanographic Institution, USA.

Figure 3.1b   Photo: Paul Snelgrove, Memorial University of Newfoundland,
Canada.

Figure 3.1c   Photo: Paul Snelgrove, Memorial University of Newfoundland,
Canada.

Figure 3.1d   Photo: Dan Fornari, Woods Hole Oceanographic Institution, USA.

Figure 3.1e   Photo: Peter Lawton, Fisheries and Oceans Canada.

Figure 3.1f   Photo: Tom Kleindinst, Woods Hole Oceanographic
Institution, USA.

Figure 3.2   Credit: Reproduced from G. D. Johnson, J. R. Paxton, T. T. Sutton,
et al., "Deep-sea mystery solved: astonishing larval transformations
and extreme sexual dimorphism unite three fish families." *Biol.
Lett.*, **5**(2009) 235–9.

Figure 3.3   Credit: author.

CHAPTER 4

Figure 4.1   Credit: Modified from N. C. Makris, P. Ratilal, D. T. Symonds, *et al.* "Fish population and behavior revealed by instantaneous continental shelf-scale imaging." *Science*, **311** (2006) 660–3. Reprinted with permission from AAAS.

Figure 4.2   Photo: Dan Costa, University of California, Santa Cruz, USA.

Figure 4.3   Credit: Reproduced from E. Vanden Berghe, K. I. Stocks, J. F. Grassle, "Data integration: The Ocean Biogeographic Information System." In A. D. McIntyre (ed.), *Life in the World's Oceans: Diversity, Distribution, and Abundance* (Oxford: Blackwell Publishing Ltd., 2010), pp. 333–53.

Figure 4.4   Credit: Ei Fujioka, Census Mapping and Visualization, Marine Geospatial Ecology Lab, Duke University, USA.

Figure 4.5   Credit: Reproduced from A. C. Bucklin, S. Nishida, S. Schnack-Schiel, *et al.*, "A Census of zooplankton of the global ocean." In A. D. McIntyre (ed.), *Life in the World's Oceans: Diversity, Distribution, and Abundance* (Oxford: Blackwell Publishing Ltd., 2010), pp. 247–65.

CHAPTER 5

Figure 5.1   Credit: used with permission of Census of Marine Life Synthesis Group, Artwork by Lianne Dunn, www.liannedunn.com.

Figure 5.2a   Photo: Mike Goebel, US-Antarctic Marine Living Resources Program.

Figure 5.2b   Photo: GoMA, Susan Ryan, University of Southern Maine, USA.

Figure 5.2c   Photo: Gary Cranitch, Queensland Museum, Australia.

Figure 5.2d   Photo: Andy Collins, National Oceanic and Atmospheric Administration, USA.

Figure 5.3a   Photo: Mike Strong and Maria-Ines Buzeta, New Brunswick, Canada.

Figure 5.3b   Photo: Humberto Bahena-Basave, El Colegio de la Frontera Sur (ECOSUR), Mexico.

Figure 5.3c   Photo: Gary Cranitch, Queensland Museum, Australia.

Figure 5.3d   Photo: Mike Strong and Maria-Ines Buzeta, New Brunswick, Canada.

Figure 5.3e   Photo: Katrin Iken, University of Alaska Fairbanks, USA.

Figure 5.4   Credit: Reproduced from L. S. Incze, P. Lawton, S. L. Ellis, and N. H. Wolff, "Biodiversity knowledge and its application for in the Gulf of Maine area." In A. D. McIntyre (ed.), *Life in the World's Oceans: Diversity, Distribution, and Abundance* (Oxford: Blackwell Publishing Ltd., 2010), pp. 43–63.

Figure 5.5a   Photo: Mike Strong and Maria-Ines Buzeta, New Brunswick, Canada.

Figure 5.5b   Photo: Gary Cranitch, Queensland Museum, Australia.

Figure 5.5c   Photo: Gary Cranitch, Queensland Museum, Australia.

Figure 5.5d   Photo: Katrin Iken, University of Alaska Fairbanks, USA.
Figure 5.5e   Photo: Brenda Konar, University of Alaska Fairbanks, USA.
Figure 5.5f   Photo: Dr. Julian Finn, Museum Victoria, Australia.

CHAPTER 6

Figure 6.1a   Photo: Russ Hopcroft, University of Alaska Fairbanks, USA.
Figure 6.1b   Photo: Kevin Raskoff, ArcOD.
Figure 6.1c   Photo: Bodil Bluhm and Katrin Iken, University of Alaska Fairbanks, USA.
Figure 6.1d   Photo: Russ Hopcroft, University of Alaska Fairbanks, USA.
Figure 6.1e   Photo: Frédéric Busson, Muséum national d'Histoire naturelle, Paris, France.
Figure 6.1f   Photo: Inigo Everson, Anglia Ruskin University, UK.
Figure 6.2   Credit: used with permission of Census of Marine Life Synthesis Group, Artwork by Lianne Dunn, www.liannedunn.com.
Figure 6.3   Credit: Data from J. M. Strugnell, M. A. Collins, and A. L. Allcock, "Molecular evolutionary relationships of the octopodid genus *Thaumeledone* (Cephalopoda : Octopodidae) from the Southern Ocean." *Antarct. Sci.*, **20**(2008) 245–51. Map produced by Ben Best, Marine Geospatial Ecology Lab, Duke University, and Huw Griffiths, British Antarctic Survey, using GeoPhyloBuilder software.
Figure 6.4a   Photo: Rolf Gradinger, University of Alaska Fairbanks, USA.
Figure 6.4b   Photo: Shawn Harper, University of Alaska Fairbanks, USA.
Figure 6.4c   Photo: Julian Gutt, Alfred Wegener Institute, University of Bremen, Germany.
Figure 6.4d   Photo: Katrin Iken, University of Alaska Fairbanks, USA.
Figure 6.4e   Photo: Julian Gutt, Alfred Wegener Institute, University of Bremen, Germany.
Figure 6.4f   Photo: Stephen Jewett, University of Alaska Fairbanks, USA.

CHAPTER 7

Figure 7.1   Credit: used with permission of Census of Marine Life Synthesis Group, Artwork by Lianne Dunn, www.liannedunn.com.
Figure 7.2a, b Credit: GeoEye satellite image. Copyright 2010. All rights reserved.
Figure 7.3a   Photo: David Patterson, used by permission of micro*scope (microscope.mbl.edu).
Figure 7.3b   Photo: Jed Fuhrman, University of Southern California, USA.
Figure 7.3c   Photo: Larry Madin, Woods Hole Oceanographic Institution, USA.
Figure 7.3d   Photo: Larry Madin, Woods Hole Oceanographic Institution, USA.

Figure 7.3e   Photo: Russ Hopcroft, University of Alaska Fairbanks © 2006, USA.
Figure 7.3f   Photo by Larry Madin, Woods Hole Oceanographic Institution, USA.
Figure 7.4a   Credit: POST 2009, Canada.
Figure 7.4b   Credit: John Healey, Vancouver Aquarium 2009, Canada.
Figure 7.4c   Credit: POST 2009, Canada.
Figure 7.5.   Credit: S. A. Shaffer, Y. Tremblay, H. Weimerskirch, *et al.*,
               "Migratory shearwaters integrate oceanic resources across the
               Pacific Ocean in an endless summer." *Proc. Natl. Acad. Sci. USA*,
               103(2006) 12799–802. Copyright (1996) National Academy
               of Sciences, USA.

CHAPTER 8

Figure 8.1a   Photo: Barbara Buge, Muséum national d'Histoire naturelle,
               Paris, France.
Figure 8.1b   Photo: Larry Madin, Woods Hole Oceanographic Institution, USA.
Figure 8.1c   Photo: © Karen Gowlett-Holmes, Commonwealth Scientific and
               Industrial Research Organization (CSIRO), Australia.
Figure 8.1d   Photo: Marco Buntzow and Paulo Corgosinho, DZMB German
               Centre for Marine Biodiversity Research, Germany.
Figure 8.1e   Photo: David Shale, www.deepseaimages.co.uk/.
Figure 8.1f   Photo: David Shale, www.deepseaimages.co.uk/.
Figure 8.2   Credit: used with permission of Census of Marine Life Synthesis
               Group, Artwork by Lianne Dunn, www.liannedunn.com.
Figure 8.3a   Photo: Canadian Scientific Submersible Facility (CSSF)/ROPOS.
               Peter Lawton, Ellen Kenchington, and Anna Metaxas, Canada.
Figure 8.3b   Photo: Bremen University/MPI; *METEOR* Expedition.
Figure 8.3c   Photo: Martin Riddle, Australian Antarctic Division.
Figure 8.3d   Photo: Craig M. Young, Oregon Institute of Marine Biology,
               and Johnson *SeaLink* Group, USA.
Figure 8.3e   Photo: Julian Gutt, Alfred Wegener Institute, University of Bremen,
               Germany.
Figure 8.3f   Credit: DTIS TAN0803 courtesy National Institute of Water
               and Atmospheric Research, New Zealand.
Figure 8.4   Credit: Reproduced from T. J. Webb, E. Vanden Berghe, R. O'Dor.
               "Biodiversity's big wet secret: the global distribution of marine
               biological records reveals chronic under-exploration of the deep
               pelagic ocean." *PLoS ONE*, 5:8 (2010), e10223, doi:10.1371/journal.
               pone.0010223.
Figure 8.5   Credit: Jesse Cleary, Census Mapping and Visualization, Marine
               Geospatial Ecology Lab, Duke University, USA.

Figure 8.6   Photo: Brigitte Ebbe and Michael Turkay, CeDAMar.
Figure 8.7a   Photo: Woods Hole Oceanographic Institution, USA.
Figure 8.7b   Photo: IFREMER/Olivier Dugornay, France.
Figure 8.7c   Credit: Arne Pallentin, National Institute of Water and
                 Atmospheric Research, New Zealand.
Figure 8.7d   Photo: University of Bremen, Germany.
Figure 8.7e   Photo: Peter Lawton, Fisheries and Oceans Canada.

CHAPTER 9

Figure 9.1   Photo: © Maria Magdalena Antczak and Andrzej Antczak.
                 Universidad Simón Bolivar, Venezuela.
Figure 9.2   Photo: The Rooms Provincial Archives, Newfoundland and
                 Labrador, VA 21-18/Robert E. Holloway, Canada.
Figure 9.3   Based on data from H. K. Lotze, H. S. Lenihan, B. J. Bourque, *et al.*
                 "Depletion, degradation, and recovery potential of estuaries and
                 coastal seas." *Science*, **312**(2006) 1806–9.
Figure 9.4   Photo: H. Blegvad, Fiskenet i Danmark.
Figure 9.5   Credit: Reproduced from L. McClenachan, "Documenting loss
                 of large trophy fish from the Florida Keys with historical
                 photographs." *Conservation Biology*, **23**(2009) 636–43.

CHAPTER 10

Figure 10.1   Credit: Reproduced from E. Vanden Berghe, K. I. Stocks, and
                 J. F. Grassle, "Data integration: the Ocean Biogeographic
                 Information System." In A. D. McIntyre (ed.), *Life in the World's
                 Oceans: Diversity, Distribution, and Abundance* (Oxford: Blackwell
                 Publishing Ltd., 2010), pp. 333–53.
Figure 10.2   Photo: Mike Johnson, earthwindow.com.

INDEX